ELECTRICAL MEASUREMENTS

Electrical Measurements

ARTHUR WHITMORE SMITH

Late Professor of Physics
University of Michigan

M. L. WIEDENBECK

Professor of Physics
University of Michigan

FIFTH EDITION

McGraw-Hill Book Company, Inc.

NEW YORK   TORONTO   LONDON

1959

PREFACE

The material for the first edition of this text was prepared more than forty years ago by the late Professor Smith. During the remainder of his life he maintained an active interest in the improvement of measuring techniques and teaching methods. The wide acceptance of this text by many universities and its use by many thousands of students in physics and engineering stand as testimony to the work to which he was devoted.

In the preface to the fourth edition Professor Smith stated, "The time has certainly come when the electron theory of electrical phenomena should be presented to all students of physics and electrical engineering. Regarding the electron tubes used in radio communication, for instance, there is no doubt that the stream of electrons through the tube continues as an electron current through the connecting wires. Since this point of view has been taken throughout the book, the drifting of electrons along the circuit is called 'an electron current,' or simply 'the current.' It is hoped that this concept, which is in accordance with the ideas of modern physics, will add a concreteness to the subject that will prove helpful in the classroom. Where the direction of this current must be indicated, an arrow is placed alongside the circuit, pointing up the potential gradient, i.e., from − to + potentials. This direction is also the direction of the electron current." This system is continued in the fifth edition.

The fifth edition has been prepared with two main goals in mind: first, to give the intermediate student a certain facility in handling work in electricity and magnetism from the theoretical as well as from the experimental standpoint; second, to prepare the student for further work in the fields of advanced electricity and electronics. At all times an attempt is made to relate the various phenomena to the basic physical laws and concepts.

All of the chapters have been revised with new material being inserted while obsolete techniques have been removed. A chapter on the treatment of errors has been added. Alternating-current circuits are discussed from the standpoint of complex notation and the various circuit theorems which have been developed in direct-current measurements are applied to the a-c circuits.

The work in electronics has been expanded to include a thorough discussion of the circuit equivalents of the vacuum tube. An elementary treatment of solid-state electronics is included along with several experiments on the use of transistors.

Problems have been added at the end of each chapter. Many of these problems are intended to stimulate the student to extend his work beyond the range covered specifically in the text.

M. L. WIEDENBECK

CONTENTS

Chapter 10. Measurement of Power 155

Chapter 11. Magnetic Effects of Electrical Currents 167

CHAPTER 1

EVALUATION OF ERRORS

1-1. Introduction. The story is often told of the athlete who also fancied himself as being a statistician. On one occasion he was anxious to swim in the ocean but had been warned that the waters were shark-infested. He decided, however, to investigate this first hand; so procuring a bucket, he proceeded to dip a pail of water from the ocean. Upon finding no sharks in the bucket he concluded that there were none in the ocean and began swimming. Of course, the late swimmer had drawn the wrong conclusion from his meager data. He had proved only that there was not an infinite number of very small sharks in the ocean.

In order to interpret the results of any quantitative measurement in an intelligent manner, an understanding and thorough evaluation of the errors are essential. By the term *error* is meant the difference between the true value of a quantity and the best measured value. While it is never possible to measure the exact value of a quantity, it is nearly always possible to give a best value and set certain limits upon the difference between the true value and the measured value. Thus when a result is given as $R = A \pm a$, what does this tell us concerning the true value of R? Are we certain that the true value of R does not differ from A by an amount greater than a, or does it mean that R is equally likely to have any of the values from $A - a$ to $A + a$? These are some of the questions which must be answered in a quantitative manner.

1-2. Systematic Errors. There are two general classes of errors which must be considered in some detail. These are systematic or persistent errors and accidental errors. Systematic errors often arise from a source which may be unknown to the observer and as such may influence the results without giving a clue as to their presence. As an example, the calibration of an ammeter may be carried out by using a standard cell as the reference. Let us suppose that the value of the emf of the cell when last measured against a primary standard was given as 1.0832 volts. Since this number is of importance in the absolute calibration of the meter, any error in this value will be reflected in the results. Although the results may be completely consistent from one set of data to another, the internal consistency will not show the presence of this accidental

1

error. As a second example, consider measurements made with a galvanometer in which the deflection of the galvanometer is measured in terms of the position of a beam of light on a calibrated scale. An error in the marking of lengths on the scale, either throughout the scale or on any portion of it, would again produce errors in the results which could go unnoticed and thereby introduce a difference between the true value and the measured value of the quantity.

1-3. Detection of Systematic Errors. Although it frequently happens that systematic errors are of such a nature that they cannot be easily corrected or evaluated, it is often possible to detect their presence or at least have greater confidence that they are absent. In general, the most straightforward method for accomplishing this is to repeat the experiment under widely differing sets of conditions. As pointed out previously, an ammeter could be calibrated with a specific standard cell, standard resistance, and potentiometer. The entire calibration could then be repeated using an entirely different set of standards as well as a different potentiometer. If the results in each case were essentially identical, one could feel reasonably confident of their correctness. If, on the other hand, the two sets of data differed, we would be aware that one or both were in error by some unknown amount.

It is of prime importance to maintain several laboratory standards of each type and to have them calibrated frequently. These should, of course, be used with the greatest of care, making sure beforehand that the measurement to be undertaken will not in itself change the standard.

It is sometimes possible to eliminate or estimate accidental errors due to instrument deficiencies even if only one instrument is available. For example, when an object is weighed with an equal arm balance, the positions of the standard weights and the object to be weighed can be interchanged to correct for unequal balance arms. Similarly when measurements are made with current or ballistic galvanometers, errors can often be detected by simply reversing the direction of deflection. As a matter of simple precaution it is always advisable when using reversible instruments to make observations in both directions.

1-4. Systematic Errors of Judgment. In addition to deficiencies in standards and in instruments, errors of the systematic type may arise from the tendency of an observer consistently to judge readings to be too high or too low. Even a trained observer who normally would have no such prejudice may find that under certain conditions of fatigue or environment, systematic errors will creep into an observation. A number of precautions can sometimes be taken to correct for or to minimize these influences.

It is almost a universal obsession of beginning students to insist that the initial reading on an instrument, such as a galvanometer, be zero.

This, of course, eliminates the process of subtracting the initial point from the end point in obtaining the value of a deflection. On the other hand, it produces an ideal situation for the introduction of errors due to prejudice. It is always safer to set the initial position at random and carry out the required subtraction. Thus, if the observer has a tendency toward high readings, the error will tend to be canceled by the subtraction. At times such effects may be taken into account by determining the magnitude and direction of such prejudices by means of controlled auxiliary experiments which can be devised. In addition, whenever it is practical, the readings should be made by two or more observers.

1-5. Accidental Errors. Let us imagine that an experiment is set up in which all systematic errors have been eliminated. A series of measurements of the same quantity is now made by the same observer and with the same apparatus in each case. It will be found that all the results will not be identical and, in fact, there may be a very large variation between the maximum and minimum values. Such a distribution of values taken under identical conditions is said to arise from accidental errors. These errors are due to a multitude of small factors which change or fluctuate from one determination to the next and are due purely to chance. The problem, then, is to evaluate such a set of data in order to ascertain the best value of the quantity being measured and in addition to set limits upon the amount by which

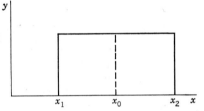

FIG. 1-1. Frequency of occurrence of events y as a function of the magnitude of the quantity x.

the magnitude of the quantity may differ from the best value while still being consistent with the data. To accomplish these things it is essential to investigate the empirical distribution of the data.

Let us assume that several measurements of a given quantity are distributed in the manner indicated in Fig. 1-1. x is the measured magnitude of the quantity, and y represents the number of occurrences of a given value. What can be said about the value of the measured quantity from these data? First of all we can say that the mean value is x_0, but we must add that any value between x_1 and x_2 has an equal probability of occurrence.

Next consider the distribution of the observed values indicated in Fig. 1-2a. This is entirely different from the distribution discussed in Fig. 1-1 and will, indeed, permit us to make a more positive assertion concerning the true value of the quantity. With this distribution x_0 is the average value of the measurements, and in addition it is also the most probable value. While it cannot be stated with certainty that the true value is

x_0, we can say that there is a 50 per cent chance that the true value lies within the limits between x_1' and x_2'. This is a more definite statement than that which was made concerning the flat distribution of Fig. 1-1. Figure 1-2b shows even greater peaking at its center. The mean and most probable value is still x_0, the maximum spread of the data still

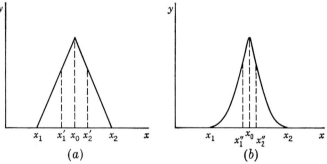

(a)   (b)

FIG. 1-2. Frequency distributions which show greater peakings at the mean value than that considered in Fig. 1-1.

extends from x_1 to x_2, but this is a more precise set of information about the quantity being measured. It is now possible to state that there is a 50 per cent chance that the true value lies in the narrow region between x_1'' and x_2''.

Under certain conditions it is possible to obtain an unsymmetrical distribution as shown in Fig. 1-3. With such an unfortunate distribution of events it becomes difficult to determine which value is the best. As previously, the mean value x_0 might be quoted. On the other hand x_a, the mode of the distribution, might be better, since it is the value which has occurred most frequently in the observations. A third possibility is x_b, the median value, since half of the observations have yielded results larger than x_b and half smaller. Such a curve usually indicates that something is drastically wrong with the experiment, and little faith can be placed in the correctness of the observations.

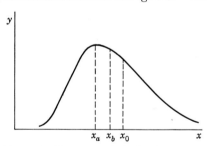

FIG. 1-3. Unsymmetrical distribution of events. x_0 is the mean value, x_a the mode, and x_b the median value.

1-6. Normal Distribution of Measurements. A normal (gaussian) distribution of values is said to exist when the accidental errors arise from a great number of very small errors which have an equal chance of being positive or negative. In practice, a smooth curve is not obtained, because each measurement covers a finite range of the quantity from x to

$x + \Delta$. For example, in reading a deflection on a galvanometer it might be decided to plot the number of events in each millimeter interval. A step curve arising from such data is shown in Fig. 1-4. When the width

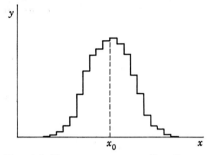

FIG. 1-4. Step curve representing the experimental data for a normal distribution.

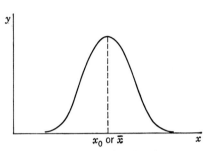

FIG. 1-5. A smooth distribution curve may replace a step curve in the limit when the interval of observation is small and the number of observations is large.

of the observed intervals Δ is small and when many observations are taken, the step curve can be replaced by a smooth curve as seen in Fig. 1-5. This would represent the normal distribution curve for the events.

1-7. Arithmetic Mean. The arithmetic mean of a series of measurements can be written as

$$\bar{x} = \frac{\sum\limits_{1}^{n} x_n}{n} \qquad (1\text{-}1)$$

The numerator is the sum of all the observed values, while the denominator is the number of observations. For a normal distribution this mean value corresponds to the peak of the distribution curve and is the most probable value which is assigned to the result.

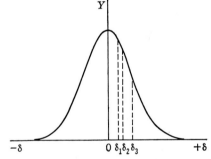

FIG. 1-6. Frequency distribution for the deviations δ.

1-8. Deviations. The term deviation is used to indicate the difference between an individual value obtained and the mean of all observations. For the type of distribution under discussion $\delta_n = x_n - \bar{x}$. A plot of the frequency distribution of the deviations can now be constructed from the normal distribution curve (Fig. 1-6). This curve can be described by the theoretical equation $Y = k\epsilon^{-h^2\delta^2}$. Y is the number of observations with a deviation from the mean equal to δ; k clearly represents the number of events which occur with zero deviation from the mean; h is the modulus of precision. The influence of the modulus of precision upon the shape of this error curve can be seen in Fig. 1-7. All the curves have been normalized to the same value of k. A high value for h represents data of high precision in which the spread of observed values is very small, while a low value of h corresponds to a less precise, flat-type distribution curve which will have greater errors associated with it.

1-9. Probability Distribution. Assume that a very large number of observations has been taken and the results yield a distribution curve

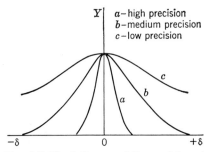

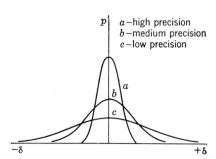

FIG. 1-7. The influence of the modulus of precision on the deviation curves. All curves have been adjusted to the same value at $\delta = 0$.

FIG. 1-8. Probability distribution curves of differing moduli of precision normalized so that the area under each is unity.

such as Fig. 1-6. We might then ask, what is the chance or probability that a given reading will occur with a deviation between δ_1 and δ_2? This probability will clearly depend upon two factors: first, the average value of Y in this region and, second, the width of the interval between δ_1 and δ_2. Had we chosen a larger interval, say between δ_1 and δ_3, the chances of finding a reading within these limits would be greatly increased.

To obtain the actual probability of finding a reading which lies within certain limits, it is, of course, necessary to normalize the distribution curve so that the total probability of finding the event with a deviation between $-\infty$ and $+\infty$ is unity. Referring to the curve in Fig. 1-7, it is noted that probability curves which have different moduli of precision cannot pass through the same point at $\delta = 0$, since the area under each curve must be the same (Fig. 1-8). This implies that k is a function of h. The equation for the probability distribution as derived by Gauss is

$$p = \frac{h}{\sqrt{\pi}} \, \epsilon^{-h^2 \delta^2} \tag{1-2}$$

The probability for finding an event in an interval between δ_1 and δ_2 then becomes

$$P = \int_{\delta_1}^{\delta_2} p \, d\delta \tag{1-3}$$

Of course, if the interval from δ_1 to δ_2 is narrow, the integral can be approximated by $P = \bar{p}(\delta_2 - \delta_1)$, in which $\bar{p}$ is the average value of the probability distribution between δ_1 and δ_2.

1-10. Precision Index. We have pointed out that the most probable or best value of a normal or gaussian distribution is obtained by taking the simple average of all the values. In addition, it has been indicated that one's confidence in the correctness of this best value is intimately connected with the sharpness of the distribution curve obtained. It would be very desirable to express this in terms of a numerical value having a specific meaning in regard to the error which may be present in the measurement. To do this we can express our results of a series of measurements in the following manner: $Q = \bar{x} \pm \varepsilon$. This tells us that the quantity Q has been measured and its best value is $\bar{x}$; however, this value is not known with certainty. The degree of uncertainty is expressed in the factor ε. The most common method for expressing this uncertainty is in terms of the *probable error*. This is defined such that there is a 50 per cent chance that the true value of Q is greater than $\bar{x} - \varepsilon$ and less than $\bar{x} + \varepsilon$. Thus a precise measurement will have a low probable error as one could already predict from the sharpness of the distribution curve.

1-11. Probable Error of a Single Measurement. There are two types of probable errors which shall be considered: first, the probable error of a single measurement from the mean and, second, the probable error of the mean. The former expresses the probability of finding the results of a *single* measurement or reading within a certain deviation E from the mean value as obtained from a large number of measurements or readings. The limits are set such that there is a 50 per cent chance that an individual reading will deviate from the mean by less than this probable value. Written in mathematical form,

$$\frac{h}{\sqrt{\pi}} \int_{-E}^{+E} \epsilon^{-h^2\delta^2}\, d\delta = \tfrac{1}{2} \tag{1-4}$$

Numerically the probable error for a single reading is $E = \pm 0.477/h$. It is sometimes possible to evaluate h from a distribution curve such as Fig. 1-6, since at the maximum $Y_0 = p_0 \Delta n = (h/\sqrt{\pi}) \Delta n$. Y_0 represents the number of observations occurring with "zero" deviation, Δ is the width of the interval of observation, and n is the total number of observations. When the measurements are too few to yield a good distribution, it is usually better to find the probable error by making use of all the individual deviations.

E is related to the deviations in the following way:

$$E = \pm 0.6745 \sqrt{\frac{\sum\limits_{1}^{n} \delta_n{}^2}{n-1}} \qquad (1\text{-}5)$$

δ_n represents the deviations of the individual measurements from the mean, and n is the number of measurements taken. The quantity $\sqrt{\sum\limits_{1}^{n} \delta_n{}^2/(n-1)}$ is called the root-mean-square deviation.

An additional curve which often finds use is shown in Fig. 1-9. This expresses the probability that an individual measurement will be found to differ from the mean by an amount less than $\pm \delta$. From the curve we see, for example, that there is an 82 per cent chance that the error will not exceed twice the probable error and a 96 per cent chance that it will not exceed $3E$.

1-12. Probable Error of the Mean Value. The question now arises as to the certainty with which we know the mean value of a set of measurements. That is, if another entirely new set of data was taken in the same manner as the first, what is the probability that the mean of the second set will fall within certain limits of the first? A second way to ask the question is, if systematic errors are absent, what is the probability that the true value lies within certain limits of the mean? This probability can again be expressed in terms of a probable error ε. There will be a 50 per cent chance that a second set of measurements of equal precision will have a mean value which does not differ from the mean of the first set by more than $\pm \varepsilon$. This probable error of the mean value is related to the probable error of a single measurement as follows:

$$\varepsilon = \pm \frac{E}{\sqrt{n}} = \pm 0.6745 \sqrt{\frac{\sum\limits_{1}^{n} \delta_n{}^2}{n(n-1)}} \qquad (1\text{-}6)[1]$$

The calculation of the various factors discussed above is illustrated with the data given in Table 1-1. These data were obtained by tossing 25 coins a total of 400 times. The number of heads was noted after each throw and is listed in column 1. The number N of occurrences of each is listed in column 2; e.g., 10 heads occurred in 38 of the 400 trials.

It is noted that h, as calculated from the maximum of the distribution curve, compares favorably with the value obtained from the root-mean-square deviation.

[1] When n is large, $n(n-1)$ may be replaced by n^2.

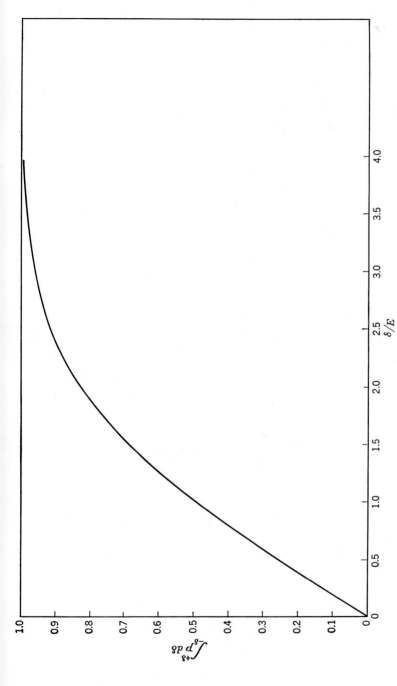

FIG. 1-9. The ordinate represents the probability that a measurement will have a deviation between $-\delta$ and $+\delta$; the abscissa is expressed in terms of the deviation considered divided by the probable error for the data.

9

TABLE 1-1. DISTRIBUTION OF EVENTS OBTAINED BY TOSSING 25 COINS
A TOTAL OF 400 TIMES

Heads	N	Nx	δ	$N\delta$	δ^2	$N\delta^2$
4	1	4	8.4925	8.4925	72.1226	72.1226
5	2	10	7.4925	14.9850	56.1376	112.2752
6	3	18	6.4925	19.4775	42.1526	126.4578
7	4	28	5.4925	21.9700	30.1676	120.6704
8	18	144	4.4925	80.8650	20.1826	363.2868
9	20	180	3.4925	69.8500	12.1976	243.9520
10	38	380	2.4925	94.7150	6.2126	236.0788
11	52	572	1.4925	77.6100	2.2276	115.8352
12	56	672	0.4925	27.5800	0.2426	13.5856
13	69	897	0.5075	35.0175	0.2576	17.3630
14	48	672	1.5075	72.3600	2.2726	109.0848
15	43	645	2.5075	107.8225	6.2876	270.3668
16	21	336	3.5075	73.6575	12.3026	258.3546
17	14	238	4.5075	63.1050	20.3176	284.4464
18	9	162	5.5075	49.5675	30.3326	272.9934
19	1	19	6.5075	6.5075	42.3476	42.3476
20	1	20	7.5075	7.5075	56.3626	56.3626

Theoretically expected value of $\bar{x} = 12.5000$

$n = 400$

Experimental value of $\bar{x} = \dfrac{\sum\limits_{1}^{n} x_n}{n} = \dfrac{4{,}997}{400} = 12.4925$

Average deviation $= \dfrac{\sum\limits_{1}^{n} \delta_n}{n} = \dfrac{831.09}{400} = 2.08$

Root-mean-square deviation $= \sqrt{\dfrac{\sum\limits_{1}^{n} \delta_n{}^2}{n}} = \sqrt{\dfrac{2715.58}{400}} = 2.61$

$Y_0 = \dfrac{h}{\sqrt{\pi}} \Delta n$ $56 + 69 = \dfrac{h}{1.77} \times 2 \times 400$ $h = 0.28$

$h = \dfrac{0.7071}{\text{rms deviation}} = \dfrac{0.7071}{2.61} = 0.27$

$E = \pm 0.6745 \text{ rms deviation} = \pm 1.76$

$\varepsilon = \pm \dfrac{E}{\sqrt{n}} = \pm \dfrac{1.76}{20} = \pm 0.088$

$\dfrac{\text{Probable error } (E)}{\text{Average deviation}} = 0.85$

Theoretically expected value of $\dfrac{\text{probable error}}{\text{average deviation}} = 0.84$

For simplicity of calculation, it is often justified to compute the average deviation, since it is closely related to the probable error E. It should also be stressed that the probable error E of a single measurement depends upon the nature of the experiment and is not greatly influenced by the number of observations. (That is, had the coins been tossed 1,000 times rather than 400 times, the value of E would be essentially unchanged.) The probable error of the mean, on the other hand, is improved as the number of observations is increased.

1-13. Propagation of Errors. The next important phase to be considered is the magnitude of the probable error which must be associated with a quantity derived from other quantities each of which may in itself have a certain probable error of its mean value. As a simple case, assume that two resistances are in series. The equivalent resistance is $\bar{R} = \bar{r}_1 + \bar{r}_2$. If the error of $\bar{r}_1$ is $\varepsilon_{\bar{r}_1}$ and that of $\bar{r}_2$ is $\varepsilon_{\bar{r}_2}$, we should not expect that the error in $\bar{R}$ will be the sum of the individual errors, since this would imply that whenever a positive deviation in $\bar{r}_1$ occurred, the deviation in $\bar{r}_2$ would also be positive. Inasmuch as these are presumed to be independent quantities, this, of course, will not happen. The error in $\bar{R}$ will, in fact, be

$$\varepsilon_{\bar{R}} = \sqrt{\varepsilon_{\bar{r}_1}{}^2 + \varepsilon_{\bar{r}_2}{}^2} \tag{1-7}$$

In a similar way, if a quantity is derived from the difference between two quantities, the errors are propagated in the same manner as for the sum of two quantities.

For the general case $\bar{R} = f(\bar{r}_1, \bar{r}_2)$ the probable error in $\bar{R}$ can be written

$$\varepsilon_{\bar{R}} = \left[\left(\frac{\partial R}{\partial \bar{r}_1} \right)^2 \varepsilon_{\bar{r}_1}{}^2 + \left(\frac{\partial R}{\partial \bar{r}_2} \right)^2 \varepsilon_{\bar{r}_2}{}^2 \right]^{1/2} \tag{1-8}$$

1-14. Weighting of Data. At times a set of data may be made up of individual values which are not equally trustworthy. It would not be proper to give such values equal importance in determining the mean value under these circumstances. The individual values can all be used, however, if they are given an importance or weight which is proportional to their trustworthiness. The mean value can then be computed:

$$\bar{x} = \frac{\displaystyle\sum_1^n w_n x_n}{\displaystyle\sum_1^n w_n} \tag{1-9}$$

w_n is the relative weight assigned to the nth reading x_n.

The assignment of weights to various readings is somewhat arbitrary, since it depends on one's judgment of the precision of individual observations.

1-15. Rejection of Data. It should be recalled that the measurements on a certain quantity represent only a small sampling of the true distribution which could be obtained from an infinite number of measurements. Since very large deviations can occur in the infinite set, they can also occur legitimately in the limited sample of this set. When they do occur, they may have a disproportionate effect on the mean value and on the probable error. The problem then arises as to what should be done about those readings which have extremely large deviations. No one doubts that something should be done, but there is a multitude of opinions as to what procedure should be followed. Some prefer to reject readings entirely if they exceed a certain limit. For example, one might arbitrarily decide to ignore all readings which have a probability of occurrence of less than $\frac{1}{20}$. The wisdom of such a rigid, sharp cutoff of the data might be questioned if the measurements consist of 20 or more readings. While the artificial cutoff may have little effect upon the mean value, it would tend to ascribe a higher precision to the measurement than is really justified. A second method which might prove useful is that of giving relative weights to these divergent measurements. This does not ignore their presence but reduces the influence which they have on the results. For example, if a value occurs which has a probability of $\frac{1}{100}$ of occurring and if 20 values are taken, one might give this value a relative weight of $\frac{1}{5}$ of the other readings.

While at times it is important to reject data, these data should always be recorded, since the too frequent occurrence of very high or very low values may sometimes indicate a defective instrument or procedure.

1-16. Significant Figures. The question as to the number of significant figures which should be given in expressing the results of a quantity is intimately connected with the error which exists in that measurement. For example, the number $\frac{4}{3}$ would not be written as 1.3333 . . . if the error in this number were ± 0.08. It would be meaningless to carry out the number to such an extent. Properly, the value should be written 1.33 ± 0.08. Similarly, when combining results of different precisions one must consider the propagated error in determining the method of expressing the results. If the precision of one part of the measurement is poor, the final result will be correspondingly poor even though the other component measurements could be made with very high precision. As an example, a resistance may be measured as V/I. Suppose that the current I is measured as 1.5 ± 0.1 amp and the voltage can be measured to six significant figures, 1.00035 ± 0.00002. Clearly the uncertainty in the value of the resistance will be no less from such a measurement than from one in which the voltage is measured to three or four significant figures.

1-17. Exercise. The beginning student is often unaware of the power of statistical methods in taking and analyzing experimental data. As an example, suppose that it is required to measure the deflection of a galvanometer on a scale which is marked in centimeters. There would be a great temptation to set the initial position at zero on the scale and then measure the final position after the deflection is produced. Suppose, now, that the deflection is actually 10.3 cm. What could be said about the observed reading under these conditions? The untrained observer would say that the reading is slightly greater than 10. The more daring type might say that it seems to be somewhere in the region between 10.2 and 10.4. Most would conclude that it is hardly worth taking more than two readings because the deflection is always the same.

By changing the procedure slightly one can not only obtain a higher accuracy but also tell the accuracy with which the deflection is known. For each reading let us set the scale at random and measure the initial position and the final position to the nearest centimeter. Some of these readings will give a value of 10 cm, while others will give 11 cm. If the reading were taken 100 times, it would be found that the value 10 was obtained in about 70 of the trials while 11 was obtained in the other 30 measurements. The average would yield a best value, and the errors could be calculated.

Experimentally determine the number of centimeters in a foot. Use a 1-ft ruler and a meter stick. Measure the length by placing the ruler on the meter stick at random. Make each measurement only to the nearest centimeter. (That is, for example, if one end is at 19.6, record it as 20; if the other end is at 49.9, record it as 50.)

Make at least 100 measurements to compute the mean, root-mean-square deviation and the probable error of the mean.

If possible exchange results with others in order to detect the existence of systematic errors.

REFERENCES

Beers, Yardley: "Introduction to the Theory of Error," Addison-Wesley Publishing Company, Reading, Mass., 1953.

Topping, J.: Errors of Observation and Their Treatment, *Monographs for Students*, The Institute of Physics, London, 1955.

Worthing, A. G., and J. Geffner: "Treatment of Experimental Data," John Wiley & Sons, Inc., New York, 1943.

PROBLEMS

1-1. The quantity $\bar{R}$ is obtained from one or more of the following measured quantities: $\bar{r}_1 \pm \varepsilon_{\bar{r}_1}$, $\bar{r}_2 \pm \varepsilon_{\bar{r}_2}$ and $\bar{r}_3 \pm \varepsilon_{\bar{r}_3}$. Compute the probable error of the mean of R for the following cases:

a. $\bar{R} = \bar{r}_1 + \bar{r}_2 + \bar{r}_3$

b. $1/\bar{R} = 1/\bar{r}_1 + 1/\bar{r}_2$

c. $\bar{R} = \bar{r}_1 \times \bar{r}_2$

d. $\bar{R} = \bar{r}_1/\bar{r}_2$

e. $\bar{R} = A\bar{r}_1$ (A is a constant)

f. $\bar{R} = \bar{r}_1{}^a$ (a is a constant)

g. $\bar{R} = \sqrt{\bar{r}_1}$

h. $\bar{R} = \sin \bar{r}_1$

i. $\bar{R} = \log \bar{r}_1$

j. $\bar{R} = \epsilon^{\bar{r}_1}$

1-2. A battery of voltage 1.52 ± 0.02 is connected across a resistance of 300 ± 3 ohms. Find the current flowing, and express its probable error.

1-3. On four occasions, 20 coins were tossed 49 times. The data are given below. Compute the mean, root-mean-square deviation, and probable error of the mean for each of the four sets of data. One set of data was taken with a two-headed coin substituted for one of the normal coins. Do the data show up this systematic error?

NUMBER OF OCCURRENCES

Number of heads	Trial 1	Trial 2	Trial 3	Trial 4
5			1	1
6	2	2	3	1
7	4	1	1	3
8	2	7	4	6
9	8	8	12	7
10	14	6	10	12
11	6	12	9	12
12	9	7	4	3
13	4	1	3	1
14		3	1	3
15		1	1	
16		1		

CHAPTER 2

REVIEW OF ELEMENTARY ELECTROSTATICS

2-1. Introduction. When beginning work in the field of electrical measurements, it is important to review the concepts of electrostatics and elementary electricity. These concepts are indispensable in gaining a deeper understanding of the quantities themselves and in applying the various principles of electrostatics to certain types of measuring instruments.

Many systems of units have been used in measuring electrical quantities as the field of electricity has evolved. The question as to the system which shall be used should not be allowed to cloud the fundamental principles. Regardless of whether a quantity or equation is expressed in the electrostatic system of units (esu) or the meter-kilogram-second system (mks), the form of the equation and its dimensions are not changed.

The mks system is rigorously adhered to in the fields of engineering. The physicist, on the other hand, does not hesitate to use the system which is most convenient or natural for the individual problem under consideration.

2-2. Electrons. The atoms of all substances are complicated structures, consisting of electrons, protons, and neutrons. The normal hydrogen atom consists of one electron, which has the smallest electrical charge (negative) that has ever been found, and a nucleus of one proton, which likewise has the smallest positive charge that has been isolated. The neutron has no charge. The atoms of other elements contain a larger number of electrons, protons, and neutrons. In certain kinds of solids some of the electrons can circulate from atom to atom, while in other solids each electron is bound rather tightly to its own atom. A substance composed of atoms of the former kind is an electrical conductor. If the atoms are of the latter kind, the substance is an insulator.

The electrons in a conductor are moving in every direction. When, in addition to this motion, there is a drift of the electrons along the conductor, this is called an "electron current."

2-3. Electronic Charge. When a piece of glass is rubbed with silk, some of the electrons are removed from the glass and are added to the silk. Just how this is done is not yet fully understood, but that does not

15

alter the fact. These electrons tend to return to their former state of equilibrium. The reason they do not do so at once is that it is almost impossible for them to move through silk or glass whereas they can move readily through copper and other metals.

It has long been the custom to call this condition of the glass "positive" electrification and the complementary condition of the silk "negative" electrification. In accordance with this established nomenclature, electrons are "negative."

Since an electron is negative, the addition of an electron to a neutral body will give it a negative charge. The addition of more electrons would increase this negative charge. The natural unit in which to express the magnitude of any charge would be the electron, but this is too small. The practical unit is defined in another way.

The protons are not readily transferred from one body to another, and the usual method of obtaining a positive charge on a body is to remove some of its electrons. The body from which the electrons are taken will be left with a positive charge.

2-4. Coulomb's Law. When a positively charged body is brought near a negatively charged body, it is found that a force of attraction exists between them. If the two bodies have like charges, either positive or negative, a force of repulsion will exist. The quantitative dependence of this electrostatic force for point charges was first investigated by Coulomb and was found to depend upon the magnitude of the charges and upon their separation in the following way:

$$F \sim \frac{q_1 q_2}{r^2} \tag{2-1}$$

This is Coulomb's law in any system of units. In the electrostatic system it is the defining equation for the quantity of charge. This is done by taking two equal point charges $q_1 = q_2$ separated by a distance of 1 cm in vacuum. If these charges now exert a mutual force of 1 dyne upon each other, the charges are said to be unit charges (1 esu of charge).[1] Thus in the esu system the constant of proportionality in Coulomb's law is unity and the law can be written:

$$F = \frac{q_1 q_2}{r^2}$$

In the mks system, the unit of force is the newton. This is the force which will produce an acceleration of 1 m/sec² when acting upon a 1-kg mass (1 newton = 10^5 dynes). The unit of charge is the coulomb,[2]

---

[1] The charge on an electron is -4.8×10^{-10} esu.

[2] The charge on an electron is -1.6×10^{-19} coulomb. One coulomb = 3×10^9 esu.

which is defined separately, and the unit of length is the meter. Since all quantities are defined independently of the equation of Coulomb's law, the constant of proportionality cannot be adjusted as in the esu system. In the mks system

$$F = \frac{1}{4\pi\epsilon_0} \frac{q_1 q_2}{r^2} \quad \text{in vacuum}$$

ϵ_0 is the permittivity of empty space. The constant $1/4\pi\epsilon_0$ has the numerical value 9×10^9.

2-5. Electrostatic Field Strength. When dealing with the forces acting upon a given electrostatic charge, it is often convenient to introduce the concept of electrostatic field strength. This may be defined as the force acting on a small, positive test charge divided by the magnitude of the test charge.

$$E = \frac{F}{\Delta q} \tag{2-2}$$

In the esu system, F is measured in dynes and Δq in esu of charge. The field strength E is therefore expressed in dynes per esu of charge. Being a force, E can be represented by a vector in the usual manner.

When the mks system is employed, field strength is expressed as newtons per coulomb.

2-6. Lines of Force. An electrostatic field can be represented by means of lines of force or flux. The magnitude of the field is given by the closeness or density of the lines, while the direction of the field in a given region is the direction of the lines of force in this region. For example, in the esu system a field strength of 1 dyne/esu would be represented by drawing one line of force per square centimeter (normal to the area) in this region.

2-7. Gauss's Theorem. If the electrostatic field in a region is produced by an ensemble of point charges, the field strength can be evaluated by applying Coulomb's law. For a continuous distribution of charge as might be encountered on a cylindrical condenser, it is often advantageous to apply a theorem set forth by Gauss. In equation form, for the esu system, this can be written

$$\iint E_n \, dS = 4\pi q \tag{2-3}$$

That is, the surface integral of the normal component of the field strength is 4π multiplied by the charge enclosed within the surface. Since the field strength can be represented by lines of flux, the surface integral can be regarded as the total flux which crosses the surface.

This theorem can be applied to the long cylindrical condenser shown in Fig. 2-1. An imaginary cylinder of radius r is constructed coaxially with the cylinders which make up the condenser. The inner cylinder of radius

r_1 has a charge of ρ esu per unit length. The outer cylinder, of course, has a similar negative charge. If the system is long compared with r_2, the lines of force will be radial.[1]

Gauss's law is written

$$(2\pi rl)E_r = 4\pi(\rho l) \qquad \text{or} \qquad E_r = \frac{\rho}{2r}$$

In the mks system Gauss's theorem can be written

$$\iint E_n \, dS = \frac{1}{\epsilon_0} q$$

E_n is in newtons per coulomb, the area is in square meters, and the charge is expressed in coulombs.

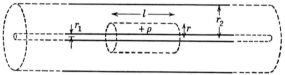

Fig. 2-1. Cylindrical condenser. The electrostatic field strength in the region between the electrodes can be computed with the aid of Gauss's theorem.

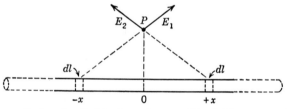

Fig. 2-2. Inner cylinder of cylindrical condenser. The field at any point is radial if the cylinder is long.

2-8. Potential Difference. If a positive test charge Δq is placed in an electrostatic field, a force will be exerted on the charge in the direction of the field $(F = E \, \Delta q)$. An external force of this magnitude will be required to move the test charge in the direction opposite the field direction. Therefore in moving the charge as described, we must exert a force through a distance and thus do a certain amount of work. Conversely, if the positive charge is moved in the direction of the field, the field will do work on it. The work per unit charge in moving from point A to point B

---

[1] The inner cylinder is represented in Fig. 2-2; the point at which the field is to be evaluated is P. An element of length dl at the position $-x$ will produce a field at P which can be represented by the vector E_1. The corresponding element at $+x$ produces a field E_2. When these two fields are combined, their horizontal components will cancel, leaving only a radial component. When the cylinder is long, each point to the left of O will have a corresponding point to the right which will cause cancellation of the horizontal components.

is the potential difference between A and B, $V_{AB} = W_{AB}/\Delta q$. In the esu system, work is expressed in ergs and charge in esu of charge. Potential difference is therefore ergs per esu.

In the mks system work is expressed in newton-meters while charge is measured in coulombs. Potential is therefore newton-meters per coulomb or joules per coulomb or the volt.

Since potential difference and field strength bear the same relationship as work and force, we can write

$$E_x = -\frac{\partial V}{\partial x} \qquad E_y = -\frac{\partial V}{\partial y} \qquad E_z = -\frac{\partial V}{\partial z}$$

$$V_{AB} = -\int_A^B E_r \, dr \tag{2-4}$$

2-9. Capacitance. For a system of electrodes (Fig. 2-3) each charged with the same magnitude of charge but of opposite sign, the capacitance can be defined as the ratio of the charge on one electrode to the potential difference between them.

$$C = \frac{Q}{V} \tag{2-5}$$

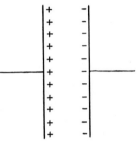

In the electrostatic system, the unit of charge is the esu and that of potential difference ergs per esu. It follows, therefore, that in this system capacitance has the unit of *centimeters*. In any system, capacitance can be reduced to a unit of length. For the mks system, the unit is called the farad and is a capacitance such that a charge of 1 coulomb will produce a potential difference of 1 volt.

Fig. 2-3. Representation of a condenser. Its capacitance is $C = Q/V$.

2-10. Energy Required to Charge a Condenser. A capacitance can be charged by carrying electrons from one electrode to the other. This can be done by mechanical or chemical means. As more and more charge is transported, the potential difference rises and makes it more difficult to carry additional charge between the plates. When the potential has risen to a value V_1, an amount of work $W = V_1 \Delta q$ will be required to carry a charge Δq from one plate to the other. The total work in charging the condenser to its final value Q will be the sum of all the work required in transporting this charge.

$$W = \int V \, dq \qquad \text{but} \qquad V = \frac{q}{C}$$

$$W = \int \frac{q}{C} \, dq = \frac{\frac{1}{2}Q^2}{C} = \frac{1}{2}CV^2 = \frac{1}{2}QV \tag{2-6}$$

In the electrostatic system W is in ergs, V in ergs per esu, q in esu, and C in centimeters. For the mks system W is expressed in joules, V in volts, Q in coulombs, C in farads.

The energy which is used in charging the condenser is not dissipated but merely stored in the electrostatic field which exists in the region of the electrodes and which has been set up by the separation of charge. It can be shown that the energy stored in a unit volume owing to an electrostatic field in a vacuum is $W = 1/8\pi \, E^2$ ergs/cu cm.

2-11. The Practical System of Units. The esu system is not convenient for dealing with the quantities encountered in current electricity. The practical system was devised to eliminate some of the inconveniences of very large and very small magnitudes which would arise in the esu system. The units of the practical system are the same as those in the mks system, differing only in the manner of definition.

2-12. Electron Current. Bodies in which the electrons can freely move from one part to another are called "electrical conductors." If electrons are added or removed at one point of a conductor, there is a movement of the electrons in the conductor to reestablish equilibrium. This flow of the electrons constitutes an electron current.

When a plate of zinc is placed in a solution of zinc sulfate, there is a strong tendency for zinc ions to go into solution, leaving electrons on the zinc plate. On the other hand, when a copper plate is placed in a copper sulfate solution, there is a great tendency for electrons in the plate to join the copper ions to form atoms of copper. The direction of these effects is noted here without stopping to consider why they are so. When these two arrangements are brought together, as in the Daniell cell, and the two plates are connected by a wire, the electrons on the zinc plate flow along the wire to the copper plate, where they meet the copper that is being deposited. In the solution, positive and negative ions travel between the electrodes. Thus there is a circulation of electrons through the circuit, the electrons in the wire drifting up the potential gradient toward the positive pole of the battery. Within the battery the energy of the chemical action at the electrodes is utilized in pulling the electrons down to the negative pole, whence they can flow up along the external circuit.

The number of electrons in a copper wire is many times greater than the number of atoms of copper. These electrons are in motion within the atoms, and a comparatively small drifting of a few of these electrons along the copper wire is sufficient to make an electron current. The wire becomes heated, and a magnetic field is set up in the space surrounding the wire. When a wire carrying an electron current is grasped by the left hand, with the thumb pointing in the direction of the electron flow, the fingers point in the direction of the magnetic field.

2-13. Electric Current. The appearance of a magnetic field around a wire has been known for well over a century (since Oersted's experiment, 1821) and has been ascribed to an electric current in the wire. The nature of this current was unknown, but a given direction was assigned to it in accordance with the magnetic field surrounding it. The current was said to flow in the direction that a right-handed screw advances when turned in the positive direction of the magnetic field.

It is now known that a current is a drifting of electrons along the wire, and since the electrons are negative in sign, they move in the direction opposite to that formerly assigned to the positive electric current.

In order to understand the language used in the older books, it is still useful to retain some of the old terms and to speak sometimes of the "positive direction" of an electric current. If the motion of the electrons is to be indicated, we can speak of the "electron current" without any confusion of terms. Since electrons are negative, either way of speaking results in the same meaning, but it is more in accord with modern ideas and the drifting of the electrons to forget the hypothetical direction of an "electric" current and to remember that the electrons leave the − side of a battery and drift up the potential gradient to the highest + position they can reach.

2-14. The Ampere and the Coulomb. As soon as one speaks of measuring a current, it is necessary to define the *unit* of current. A current has a magnetic effect around the wire carrying it, and when placed near a magnet or another current the wire experiences a push tending to move it sidewise across the magnetic field in which it finds itself.

The unit of current is named an "ampere."[1] This name has long been used; it was officially adopted in 1881 by the Paris Congress of Electricians.

A current of 1 amp is established as follows. Two parallel currents, I and I', in the same direction are urged toward each other with a force proportional to the product of these currents. This apparent attraction is because each current is in the magnetic field of the other, and it tends to move directly across this field, *i.e.*, toward the other wire. When the wires are 1 m apart in air (strictly, in vacuum) and the same current $(I = I')$ in each wire is large enough to give a push on each wire of 2×10^{-7} newton per meter length of the wire, this current is 1 amp.

In framing the definition of any quantity, it is necessary to avoid using the name of any unit that has not been defined previously. Since in this chapter no magnetic unit has thus far been defined, it is necessary to state the value of the ampere without mentioning any other magnetic quantity. This is done as follows:

[1] André Marie Ampère (1775–1836), French physicist.

Definition.   One ampere is the current in each of two long parallel wires placed one meter apart in vacuum when the force per unit length acting on each wire is 2×10^{-7} *newton per meter.*

The number 2×10^{-7} occurs because of the units employed (meters, newtons) and the omission of the unnecessary reference to a magnet. This definition leads directly to the practical determination of a current in amperes by weighing on a sensitive balance the force between two coils when carrying this current, as is done with extreme accuracy at the National Bureau of Standards.[1]

The coulomb is the unit of quantity of charge in the mks and practical systems of units.   Since current is the rate of flow of charge ($i = dq/dt$), charge can be written $q = \int i \, dt$.   The coulomb can thus be defined as that charge passing a point in one second when the current is one ampere.

2-15. The Ohm.   The amount of current that flows in a given circuit is determined by the resistance of the circuit as much as by the source of electromotive force.   The wire carrying the current becomes heated, and the greater the resistance, the more heat is given off each second for the same value of the current.   This expenditure of energy in the wire is used to define the unit of resistance, which is called an "ohm."[2]

The rate at which energy is thus expended by a current of I amp in a resistance of R ohms (*i.e.*, the power) is

$$P = RI^2 \qquad \text{watts} \qquad (2\text{-}7)$$

This relation shows at once just how much resistance is to be taken as 1 ohm.   This name has long been used; it was reaffirmed by the Paris Congress of Electricians in 1881.

Definition.   One ohm is that amount of resistance in which the power expended is one watt when the current is one ampere.

2-16. How Much Is 1 Watt?   To answer this question it is necessary to make a brief survey of the units that are used in mechanics.

The meter is the standard of length.   All other lengths, whether miles, inches, or centimeters, are calibrated in terms of the standard meter.

The kilogram is the fundamental standard of mass.   All other units, whether pounds or grams, are calibrated in terms of the standard kilogram, which in the United States is kept at the National Bureau of Standards.

The Second.   The unit of time in physics, and for most measurements, is the mean solar second.   Astronomers use the sidereal second, which is slightly shorter.

---

[1] Curtis, Driscoll, and Critchfield, *J. Research Natl. Bur. Standards*, vol. 28, p. 133, 1942.

[2] Georg Simon Ohm (1787–1854), German physicist.

Derived Units. From these three fundamental units, the unit of velocity is a meter per second, and the unit of acceleration is a *meter per second* per second. The force that will give this acceleration to a mass of one kilogram is called a "newton."[1] This name for the unit of force was adopted in 1938 by the International Electrotechnical Commission at their meeting in Torquay. A force of 1 newton will just about lift a weight of 102 g.

Power is expended when a force is applied to a moving body, as when a boy draws a sled or a horse pulls a cart. If the cart is stuck fast and cannot be moved, no power is expended upon it, although the force may be great. Power is measured by the product of the applied force and the velocity in the direction of this force. The power expended by a force of one newton when it is lifting a weight with a constant velocity of one meter per second is one *watt.*[2] This name for the unit of power was adopted by the International Congress of Electricians at Chicago in 1893.

Other Units. In the illustration, just given, of a force lifting a weight, the power remains constant whether the operation lasts for a minute or an hour. But the amount of *energy* expended, or the *work* done, increases as time goes on. When the power is one watt, the energy expended each second is one *joule.*[3] This name for the unit of energy was also adopted by the International Congress of Electricians at Chicago in 1893.

2-17. Difference of Potential, EMF, Fall of Potential. The difference of potential between two points is that difference in condition which produces an electron current from one point to the other as soon as they are connected by a conductor. Every battery or other electric generator possesses a certain ability to maintain a difference of potential between its terminals, and, therefore, the ability to drive a continuous current. This difference of potential produced by a cell or other generator, and which may be considered as the cause of the current, is called "electromotive force." It must be remembered that this quantity is not a force in the sense that this name is used in physics. In order to avoid using the word "force," it is commonly called "emf."

When a current flows through a conductor, there is a difference of potential between any two points on the conductor. This difference of potential is greater the farther apart the points are taken, and as the change is gradual, it is usually called a "fall of potential." It might also be called a "rise of potential," since, when *following* the electron current through a resistance, one is led to points that are more strongly + in potential. It can always be expressed by the formula RI, where R is the resistance of the conductor, or conductors, under consideration.

[1] Sir Isaac Newton (1642–1727), English natural philosopher.
[2] James Watt (1736–1819), Scottish engineer.
[3] James Prescott Joule (1818–1889), English physicist.

This apparent duplication of names may at first appear unnecessary, but the corresponding ideas are quite distinct, and the correct use of the proper term will add conciseness to one's thinking and speaking. Thus we have the emf of a battery; the fall of potential along a conductor; and the more general and broader term, "difference of potential," which includes both of the above as well as some others for which no special names are used.

2-18. Ohm's Law. *The electron current that flows through any conductor is directly proportional to the potential difference between its terminals.* This statement was first formulated in 1827 by Ohm, as the result of many experiments and measurements, and it is known as Ohm's law. It is usually written

$$V = RI \qquad \text{or} \qquad I = \frac{V}{R} \tag{2-8}$$

where V denotes the potential difference over the circuit through which is flowing the current I. The factor R is called the "resistance" of the conductor, and its value depends only upon the dimensions and material of the wire and its temperature. It is entirely independent of V and I.

This relation holds equally well whether the entire circuit is considered or only a portion of such circuit is taken. In the former case the law states that the current through the circuit is equal to the total emf in the circuit divided by the resistance of the entire circuit, including that of the battery and the connecting wires. When applied to a single conductor AB, the law states that the current flowing through the conductor is equal to the fall of potential between A and B divided by the resistance of this part of the circuit.

2-19. The Volt. Now that the value of an ampere (Sec. 2-14) and the value of an ohm (Sec. 2-15) have been defined, it follows that the value to be used as 1 volt[1] is given by Ohm's law.

Definition. One volt is the difference of potential that, steadily applied to a conductor whose resistance is one ohm, will produce a current of one ampere.

The name "volt" has long been used; it was reaffirmed by the Paris Congress of Electricians in 1881.

From the definition of the ohm it also follows that the power necessary to maintain a current of I amp under a difference of potential of E volts is

$$P = EI \qquad \text{watts} \tag{2-9}$$

This leads to a second way of stating the definition for a volt. Since $E = P/I$, we can say,

One volt is the difference of potential that will maintain a current when the power expended is one watt per ampere.

---

[1] Alessandro Volta (1745–1827), Italian physicist.

REFERENCES

Peck, E. R.: "Electricity and Magnetism," McGraw-Hill Book Company, Inc., New York, 1953.
Harnwell, Gaylord P.: "Principles of Electricity and Electromagnetism," McGraw-Hill Book Company, Inc., New York, 1949.
Fowler, R. G.: "Introduction to Electric Theory," Addison-Wesley Publishing Company, Reading, Mass., 1953.

PROBLEMS

2-1. An array of point charges is set up as follows: $+50$ esu at $x = -3$, $y = 0$, $+100$ esu at $x = 0$, $y = 0$, -50 esu at $x = +3$, $y = 0$.
 a. Find the electrostatic field strength, in both magnitude and direction at the point $x = 0$, $y = 4$.
 b. Find the electrostatic field strength at the point $x = 1$, $y = 0$.
 c. Find the electrostatic field strength at the point $x = 7$, $y = 0$.
2-2. Prove that the electrostatic field strength inside a charged, hollow, conducting sphere is zero.
2-3. Give arguments to show that the potential inside a charged, hollow, conducting sphere is a constant equal to that of the surface of the sphere.
2-4. Prove that all the charge on a conductor resides on the surface if there are no currents flowing in the conductor. (*Hint:* First show that the field everywhere inside the conductor is zero, then apply Gauss's theorem to an imaginary surface constructed entirely within the metal.)
2-5. Using Gauss's theorem, find the electrostatic field strength in the region between two concentric metal spheres of radii r_1 and r_2. The inner sphere has a charge $+q$ esu; the outer sphere is grounded and therefore has a charge of $-q$.
2-6. From the results of Prob. 2-5, compute the potential difference between these spheres.
2-7. What is the capacitance of the spherical condenser in Prob. 2-5?
2-8. Neglecting end effects, prove that the capacitance, in the electrostatic system, of a parallel plate condenser in vacuum is $C = A/4\pi d$. (A is the area of one of the plates; d is the separation between the plates.)
2-9. Find the capacitance per unit length of a Geiger counter. (This is essentially a cylindrical condenser.) The radius of the inner cylindrical wire is 0.002 cm; the outer cylinder has a radius of 1.5 cm. Neglect end effects. Also express your results in micromicrofarads.
2-10. In the mks system, electrostatic field strength is expressed as newtons per coulomb. Show that this is identical with the unit volt per meter.
2-11. From the relationship between the dyne and newton, the esu of charge and the coulomb, and the centimeter and the meter, find the constant of proportionality K in Coulomb's law when the various quantities are expressed in the mks system.

$$F(\text{newton}) = K \frac{q_1 q_2 (\text{coulomb}^2)}{r^2 (\text{m}^2)}$$

2-12. Find the stress (force per unit area) on each of the spheres in Prob. 2-5.

CHAPTER 3

AMMETER AND VOLTMETER METHODS

3-1. Introduction. The magnitude of a quantity can be measured either by a direct comparison with a standard or by effects which the quantity may produce. An example of a direct measurement is the determination of the length of a rod by comparing it with a calibrated meter stick or the measurement of a potential difference by comparison with a standard cell. Such measurements are not always possible in electricity, and it is often necessary to measure a quantity by other means. A quantity of charge, when passed through a silver nitrate solution, will electroplate atoms of silver. This gives a correspondence between the weight of silver deposited and the charge passed through the solution, thus providing a method for measuring electrical charge.

When a current passes through a wire, two effects are easily noticeable: The wire becomes hot, and a magnetic field is set up in the space surrounding the wire. These effects provide methods by which the current can be measured. The heating of a wire will cause it to increase in length, so that a correspondence exists between the length of a wire and the current passing through it. Such hot-wire ammeters have been built, but their shortcomings are numerous.

3-2. The Portable Ammeter. The portable ammeter measures a current by means of the magnetic effects produced by this current. To understand the operation of the ammeter it is necessary to introduce some of the principles of magnetism which will be discussed in greater detail in the later chapters. It is assumed that the reader is acquainted with the elementary concepts of the magnetic field as presented in elementary physics.

If a current element (an electron current i flowing in a wire of length dl) is placed in a magnetic field of strength B, a force will be exerted upon the current element which is at right angles to the direction of the field and also to the wire. Figure 3-1a shows the direction of the field B and the wire dl. The force is perpendicular to and out of the page. A top view is indicated in Fig. 3-1b. It will be noted that two magnetic fields exist and interact. These are the external field B and the field produced by the current in dl. The resultant magnetic field in the region of the current element is altered, being increased on one side of the wire (since both fields are in the same direction) and decreased on the opposite side where the fields are opposed. The force acting on the current element is always

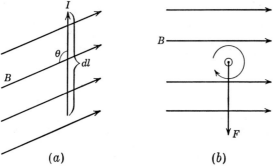

(a) (b)

FIG. 3-1. (a) Current element in a magnetic field. The electron current is indicated by the arrow. (b) Top view showing the external field B and the field produced by the current in the wire.

such as to tend to move it from the stronger to the weaker region of the field. In any system of units the force can be written

$$F = KBI(dl) \sin \theta \tag{3-1}$$

The constant of proportionality K will be determined by the system used. θ is the angle between the direction of the field and dl. This equation furnishes the basic principle underlying nearly all ammeters, that is, the direct proportionality between the current in a wire and the force on

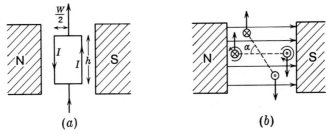

(a) (b)

FIG. 3-2. (a) Rectangular coil of wire suspended between the poles of a magnet. (b) Top view showing the directions of the fields and forces.

that wire if all other conditions (B and the geometry) remain fixed. Since it is rather difficult to measure a small force with a high degree of accuracy, an ammeter makes use of an arrangement whereby the torque produced by this force is measured.

A simple system is shown in Fig. 3-2. A rectangular coil of N turns of wire is suspended in the uniform magnetic field of strength B. As seen in Fig. 3-2a, electron current flows upward in the right side and downward in the left. Figure 3-2b is a top view of the same system showing the directions of B and the field produced by the current I. The two forces produce a torque tending to rotate the coil in a clockwise direction (looking downward). If the coil is supported by a spring or suspension arrangement, it will rotate about its axis until the torque produced by the magnetic force is just counterbalanced by the twisting of the spring or suspension. If this equilibrium is reached when the coil has turned through an angle α, we can write

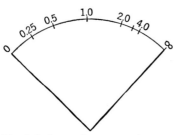

FIG. 3-3. Approximate marking of a scale for the type of instrument shown in Fig. 3-2.

$$2K(NBIh)\left(\frac{W}{2}\cos\alpha\right) = \tau\alpha \qquad (3\text{-}2)$$

$NBIh$ is the force on one side of the coil; $W/2 \cos \alpha$ is the perpendicular distance from the direction of the force to the center of rotation. Since there are two such forces acting to produce rotation in the same direction, the total torque is given by the left side of the equation. τ is the torsional constant of the suspension system and represents the torque required to twist it through a unit angle. α is the angle through which it has been twisted. Since (hW) is the area of the coil,

$$I = \frac{\tau}{KBAN}\frac{\alpha}{\cos\alpha}$$

The current is now related to the angle through which the coil turns. A pointer could be attached to the coil, and a scale marked off to indicate the current in amperes. Such a scale would not be linear (Fig. 3-3), since the current is proportional to the angle α divided by the cosine of α. It is far more desirable to have an instrument in which I and α are directly proportional. This can be accomplished by shaping the direction of the magnetic field properly as seen in Fig. 3-4. The pole faces of the magnet are given a cylindrical shape, and a cylinder of iron is fixed between them. This causes the magnetic field to be radial, so that the torque is always $2K(NBIh)(W/2)$ and the deflection of the coil is directly proportional to the current.

$$I = \frac{\tau\alpha}{KNBA} \qquad (3\text{-}3)$$

In ammeters for measuring large currents, a low-resistance shunt is placed in parallel with the moving coil to allow only a moderate current through the latter. The scale is then graduated to read the value of the large total current through both the coil and its shunt. This arrangement is entirely similar in principle to the voltmeter and shunt described in Sec. 3-24, the moving-coil system acting as a sensitive voltmeter.

3-3. Use of an Ammeter. The portable ammeter is a good and accurate instrument for the measurement of electron current. As it is very delicate and sensitive, it must always be handled with care. Mechanical shocks or jars will injure the jeweled bearings, and too large a current through it will wrench the movable coil and bend the delicate pointer, even if the instrument is not burned out thereby.

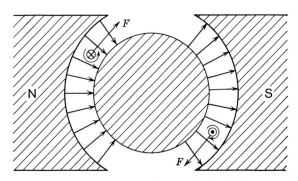

Fig. 3-4. Rectangular coil suspended in a radial magnetic field. The torque produced by the current in the coil is now independent of α.

When it is desired to use the ammeter for the measurement of current, it is connected in series with the rest of the circuit, and therefore the entire current passes through the instrument. Great care should always be exercised never to allow a larger current to flow through an ammeter than it is intended to carry. It is always best to have a key in the circuit, and while keeping the eye on the needle of the ammeter, to tap the key gently, thus closing it for a fraction of a second only. If the needle does not move very far, the key can be held down for a longer time. If it is then seen that the needle will remain on the scale, the key can be held down until the needle comes to rest. Behind the needle is a strip of mirror, and by placing the eye in such a position that the image of the needle is hidden by the needle itself, the error due to parallax in reading the scale can be avoided.

The scales of these instruments are graduated to read the current directly in amperes. Sometimes the pointer does not stand at the zero of the scale when no current is flowing. When this is the case, the position of rest should be carefully noted and the observed reading corrected accordingly.

3-4. Laws of Electron Currents. For this exercise, a dry cell, a coil of several ohms' resistance, a key, and the ammeter are connected in series, i.e., one after the other to form a single and continuous circuit.

The electrons flow from the negative, or zinc, pole of the cell out into the external circuit. In order to read the ammeter properly, it should be connected into the circuit so that the electrons will enter it at the post marked − and leave the ammeter at the post marked +. Measure and record the value of the current at different points along this circuit, to determine whether the current has the same value throughout its path or is smaller after passing through the resistances. Next add the remaining coils to the circuit, keeping them all in series, and note the value of the

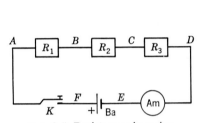

Fig. 3-5. Resistances in series.

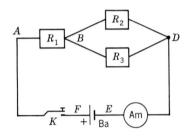

Fig. 3-6. Resistances R_3 and R_2 in parallel.

current at the same points as before. State in your own words the effect of adding resistance to the circuit.

Remove one of the coils and connect it in parallel with one of those still remaining in the circuit, *i.e.*, so that the current in the main circuit will divide, a part going through each of the two coils in parallel. Measure the part of the current through each coil; also the main current. Could you have foretold the value of the latter without measuring it?

This exercise should show that the only thing in common to coils in *series* is the value of the *current* passing through them. For this reason an ammeter is always joined in series with the circuit in which the value of the current is desired. Record the readings of the ammeter in a neat tabular form similar to the following:

To Measure the Current along an Electrical Circuit

Ammeter zero	Ammeter reading	Current	Point at which current is measured

3-5. The Portable Voltmeter.
The construction of a voltmeter is the same as that of an ammeter, except that instead of a shunt in parallel with

the moving coil, there is a high resistance in series with the coil. The current which will then flow through this instrument of high resistance depends upon the emf applied to its terminals, and the numbers written on the scale are not the values of the currents in the coil but are the corresponding values of the fall of potential over the resistance of the voltmeter. Therefore it is sometimes said that a voltmeter is really only a sensitive ammeter measuring the current through a fixed high resistance, while an ammeter is really only a sensitive voltmeter measuring the fall of potential over a low resistance.

3-6. Use of a Voltmeter. The portable voltmeter is a good and accurate instrument for the measurement of a fall of potential. Like the ammeter, it is very delicate and sensitive and must always be handled with care. Mechanical shocks or jars will injure the jeweled bearings, and too large a current through it will wrench the movable coil and bend the delicate pointer, even if the emf is not much larger than that intended to be measured by the instrument.

Since a voltmeter is always used to measure the difference of potential between two points, it is not put into the circuit like an ammeter, but the two binding posts of the voltmeter are connected directly to the two points whose difference of potential is desired. The voltmeter thus forms a shunt circuit between the two points. There is a current through the voltmeter proportional to the difference of potential between the two points to which it is joined. This current passing through the movable coil of the instrument deflects the pointer over the scale, but the latter is graduated to read not the current but the number of volts between the two binding posts of the voltmeter.

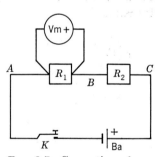

FIG. 3-7. Connections for a voltmeter.

3-7. Fall of Potential in an Electrical Circuit. For this exercise join a cell, two coils of several ohms' resistance, and a key in series. With the voltmeter measure the fall of potential over each coil, also over both together. Add a third coil and measure the fall of potential over each, also over all. Note where these last readings are the same as they were before and where they are different. Add a second cell and repeat the above readings. Note changes.

Join two of the coils in parallel, thus forming a divided circuit and allowing a part of the current to flow through each branch (see Fig. 3-6). Measure the fall of potential over each branch. Add a third coil in parallel with the other two and again measure the fall of potential over each.

This exercise should show, especially in connection with the preceding one, that the only thing common to several circuits in *parallel* is that each one has the same *fall of potential*. Hence a voltmeter is always joined in parallel with the coil over which the fall of potential is desired. The voltmeter indicates the fall of potential over itself; and if it forms one of the parallel circuits, the fall of potential over each one is the same as that indicated by the voltmeter.

Record the voltmeter readings as below:

FALL OF POTENTIAL IN AN ELECTRICAL CIRCUIT

Voltmeter		Fall of potential	Position of voltmeter
Zero	Reading		

3-8. Measurement of Resistance by Ammeter and Voltmeter. *First Method.*

To determine the resistance of a conductor, it is necessary only to measure with an ammeter the current flowing through it and to measure with a voltmeter the difference of potential between its terminals. In case the current is at all variable, the two instruments must be read at the same time, for Ohm's law applies only to simultaneous values of the current and voltage. The conductor, whose resistance R is to be measured, is joined in series with an ammeter Am, a key, a battery, and sufficient auxiliary resistance to keep the current from being too large. The electron current should leave the ammeter at the post marked $+$. Keeping the eye fixed on the needle of the ammeter, close the key for a fraction of a second. If the deflection is in the right direction and is not too large, the key can be closed again and the value of the current read from the scale of the ammeter.

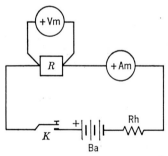

FIG. 3-8. To measure a small resistance R.

Should the current be too large, the auxiliary resistance can be increased until the current is reduced to the desired value.

To measure the fall of potential over the conductor, its two terminals are joined to the binding posts of the voltmeter by means of additional wires. That terminal of the resistance at which the electron current

leaves should be joined to the voltmeter post marked +. Close the key
for an instant as before, keeping the eye on the voltmeter needle, and if
the deflection is in the right direction and not too large, the reading of the
voltmeter can be taken.

Now close the key and record simultaneous readings of the ammeter
and the voltmeter. Do this several times, changing the current slightly
by means of the auxiliary resistance before each set of readings. Compute the resistance of R from each set of readings by means of the formula

$$R = \frac{V}{I}$$

where V and I are the voltmeter and ammeter readings, corrected for the
zero readings. The mean of these results will be the approximate value of
the resistance.

A more exact value of R can be obtained by correcting the current as
measured by the ammeter for the small current i that flows through the
voltmeter. The current through R is, strictly, not I, but $I - i$. This,
then, gives

$$R = \frac{V}{I - i}$$

The value of the current i through the voltmeter can be computed. If
the resistance of the voltmeter is S
ohms,

$$i = \frac{V}{S}$$

3-9. Measurement of Resistance by Ammeter and Voltmeter. *Second Method.*

This method differs from
the first method by the position of the
voltmeter. In the first method the
ammeter measured both the current
through R and the small current

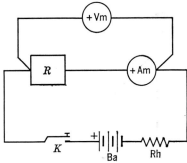

Fig. 3-9. To measure a large resistance R.

through the voltmeter, and therefore
its readings were somewhat too high.

If the connections are made as shown in Fig. 3-9, this error is avoided,
as the current now passing through R is strictly the same as that measured by the ammeter. The voltmeter, however, now measures the fall of
potential over both R and the ammeter, and therefore the resistances of
both are measured together. The resistance of R is then found by subtracting the resistance of the ammeter from the measured amount. The
formula then becomes

$$R = \frac{V}{I} - A$$

where A is the resistance of the ammeter.

This correction is easily made, and therefore this is the preferable method, except for very small resistances.

Measure the resistance of two coils and check results by also measuring them in series and in parallel. The measured resistances should be compared with the values computed from the formula—for series, $R = R_1 + R_2$, and for parallel,

$$R = \frac{R_1 R_2}{R_1 + R_2}$$

Record the readings as follows:

To Measure the Resistance of . . .

Coil measured	Ammeter reading	I	Voltmeter reading	V	Resistance	
					$\frac{V}{I}$	Corrected

3-10. Graphical Solution for Resistances in Parallel.

The equivalent resistance of two resistances in parallel is easily determined by a simple geometrical construction. Let two lines a and b be drawn at right angles to a base line PQ and at a convenient distance m apart. Let the lengths of these lines represent the two resistances on a convenient scale. Join the top of each line to the bottom of the other, as shown by the broken lines. From the intersection of these diagonals draw the line c, also perpendicular to the base line. The length of c gives the equivalent resistance of the parallel circuits on the same scale that was used in drawing a and b.

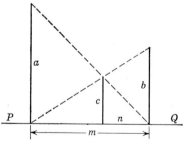

FIG. 3-10. The resistance of a and b in parallel is shown by c.

This is easily shown. From similar triangles,

$$\frac{c}{a} = \frac{n}{m}$$

and also
$$\frac{c}{b} = \frac{m-n}{m} = 1 - \frac{n}{m} = 1 - \frac{c}{a}$$

from which
$$c = \frac{ab}{a+b}$$

This diagram will thus show quickly the effect of connecting any two resistances in parallel.

3-11. Comparison of the Two Methods. In each of the preceding methods for measuring resistance by means of an ammeter and a voltmeter, it is necessary to apply a correction to the observed readings in order to obtain the true value of the resistance being measured. It is the object of this section to inquire under what conditions these corrections are a minimum. In order to compare the two correction terms with each other, they will both be expressed in the form of factors by which the observed values are multiplied to obtain the true values.

In the first method the correction is applied to the ammeter reading, the true current through R being $I' - i$. The true value of the resistance is, then,

$$R = \frac{V'}{I' - i} = \frac{V'}{I' - \dfrac{V'}{S}} = \frac{V'}{I'\left(1 - \dfrac{V'}{I'S}\right)} = \frac{R'}{1 - \dfrac{R'}{S}} = R'\left(1 + \frac{R'}{S}\right)$$

approximately, where S is the resistance of the voltmeter and R' is written for V'/I', the uncorrected value of R.

In the second method the value of R is given by

$$R = \frac{V''}{I''} - A = R'' - A = R''\left(1 - \frac{A}{R''}\right)$$

where A is the resistance of the ammeter, and R'' is written for the uncorrected value of R in this method.

In the first method the correction factor is nearly unity when the resistance being measured is small. Other things being equal, a voltmeter having a high resistance S is better than a low-resistance instrument. In the second method the correction factor is near unity when the resistance being measured is large. The smaller the resistance of the ammeter, the better. Either method, thus corrected, should give the correct resistance of R.

3-12. To Find the Best Arrangement for Measuring Resistance with an Ammeter and a Voltmeter. The division point between the two methods is at the resistance for which the correction factors R'/S and A/R'' are equal. For this resistance neither of the corrections is very large, and it is nearly true to write

$$R' = R'' = R$$

and

$$\frac{R}{S} = \frac{A}{R}$$

The division point is, then, at

$$R = \sqrt{AS}$$

This resistance is not a fixed value; it depends upon the resistances of the ammeter and the voltmeter that are used.

For resistances smaller than this, the first method has the smaller correction. For larger resistances, the second method should be used. Whether the correction is applied or not, it should always be made as small as possible. If the proper method is selected, the correction will almost never be as large as 1 per cent.

The curves in Fig. 3-11 show the factor by which R' or R'' must be multiplied to give the true resistance when a 200-ohm voltmeter and a 0.02-ohm ammeter are used. It is seen that with a resistance of 2 ohms the corrections are equally large for each method, one giving uncorrected results that are 1 per cent too high, the other giving uncorrected results that are 1 per cent too low.

In measurements of this kind a larger error may be introduced by neglecting the temperature of the wire that is being measured. The resistance of a copper wire increases about 2 per cent for each 5°C rise in temperature, and unless the temperature of the wire is known more closely than 2°C, there is not much use in measuring the resistance more closely than 1 per cent.

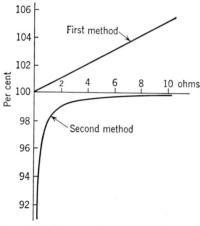

FIG. 3-11. Showing the errors in the uncorrected values given by V/I in the two methods of measuring resistance.

3-13. Relation between Terminal Voltage and Current. The battery or emf to be examined is joined in series with a variable resistance, a key, and an ammeter, as shown in Fig. 3-12. By using various values of R, the current can be varied throughout its possible range, but R should not be reduced beyond the point that gives a current as large as can be measured with the ammeter. A voltmeter is joined in parallel with the resistance and key. When the key is closed, the current drawn from the cell is

$$I = \frac{E}{R' + r'}$$

where R' is the combined resistance of R and the voltmeter in parallel, and r' includes not only the resistance of the battery but also that of the ammeter. Clearing of fractions,

$$E = R'I + r'I$$

The voltmeter measures $R'I$. Denoting this by E',

$$E = E' + r'I \tag{3-4}$$

The term $r'I$ is the fall of potential over the internal resistance of the battery (and the resistance of the ammeter if connected as in Fig. 3-12). E' ($= R'I$) is the fall of potential over the entire external part of the circuit. Various names have been applied to this term E', such as "terminal emf," "terminal potential difference," "pole potential," "available emf," etc. It is that part of the total emf of the cell which is available for doing useful work. Its value, from Eq. (3-4), is

$$E' = E - Ir' \tag{3-5}$$

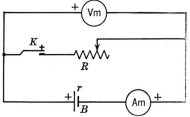

Fig. 3-12. To measure the internal resistance of the battery B. Also to measure the terminal voltage corresponding to the current drawn from the battery and the power expended in R.

and from this it appears that the terminal voltage is less as the current becomes larger.

The key should be kept closed as little as possible to avoid unnecessary polarization of the battery.

The observations can be recorded as follows:

RELATION BETWEEN AVAILABLE EMF AND CURRENT FOR A . . . CELL

Ammeter		I, amperes	Voltmeter		E', volts
Zero	Reading		Zero	Reading	

3-14. Plotting the Curve. From the values of E' and I a curve can be plotted, as in Fig. 3-13. The currents should have been chosen so that the plotted points will be well distributed along this curve. The curve shows that the available emf or terminal voltage continually decreases as more current is drawn from the source, and it will decrease to zero if the current is sufficiently increased. This is true whether the current is supplied by a battery, a dynamo, or any other source, but it is not always safe to allow this current to flow if its value would be too large.

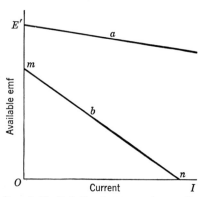

FIG. 3-13. Relation between the available emf and the current drawn from the battery. Curve a shows a low resistance in the battery (a). Curve b indicates that the battery (b) has a high internal resistance (and also a smaller emf).

For the sake of comparison, curves should be drawn for different types of cells using at least one cell of low internal resistance. Determine from the curves the maximum current that each cell can furnish. Find one reason why an ammeter should not be joined to the poles of a cell, as is done with a voltmeter.

3-15. Maximum Current from a Battery. The current from a given battery depends upon the resistance in the circuit and by Ohm's law is

$$I = \frac{E}{R' + r'}$$

The smaller the resistance of the circuit, the larger the current. The limit comes when the external resistance R' has been reduced to zero, and then

$$I = \frac{E}{r'}$$

is the maximum current that can be drawn from the battery when it is connected with an ammeter as shown in Fig. 3-12.

The value of this maximum current is shown by the intercept on the horizontal axis in Fig. 3-13. For curve b this intercept On is shown in the figure. For curve a the value of the intercept must be estimated from the slope of the portion of the curve that is shown in the figure.

With a storage battery or a dynamo, where r is very small, this maximum current may be dangerously large. No attempt should ever be made to measure the current without first knowing approximately what its value may be, or else starting with considerable resistance R in the external circuit and cautiously reducing it while the ammeter is watched as the current increases. A single storage cell with internal resistance $r = 0.001$ ohm will supply over 1,000 amp on short circuit. If the ammeter is designed to measure as large a current as this, it will not be injured. But it would be rather hard on the cell and ruinous to an ammeter of smaller range.

3-16. Maximum Voltage. The values of the terminal voltage shown in Fig. 3-13 are given by the relation

$$E' = E - r'I$$

For the particular point on this curve corresponding to $I = 0$, the ordinate is

$$E' = E$$

Therefore the value of the full emf of the battery b is given by the intercept Om on the vertical axis in Fig. 3-13.

3-17. Internal Resistance of a Battery. The internal resistance of a battery or other source of current is shown by the available emf curve, like a in Fig. 3-13, and its value is readily determined from the curve. For one of the smaller values of the current I_1, the corresponding available emf is

$$E_1' = E - r'I_1 \tag{3-6}$$

At a second point on this curve, where E_2' and I_2 are the corresponding values,

$$E_2' = E - r'I_2 \tag{3-7}$$

The unknown emf E of the battery can be eliminated by subtracting Eq. (3-7) from Eq. (3-6), giving

$$E_1' - E_2' = r'I_2 - r'I_1 \tag{3-8}$$

The internal resistance of the source, with its connections up to the points where the voltmeter (Fig. 3-12) is joined to the main circuit, has been denoted by r' in these equations, and solving Eq. (3-8) gives

$$r' = \frac{E_1' - E_2'}{I_2 - I_1} \tag{3-9}$$

Stated in words, this result is

$$\text{Resistance} = \frac{\text{change in potential difference}}{\text{change in current}}$$

which is a general statement of Ohm's law.

Thus by using two values of the terminal voltage, corresponding to two widely different values of the current, it is not necessary to use Eq. (3-6) and measure the emf of the battery, which, as shown by Fig. 3-13, can hardly be done with a voltmeter unless r is very small.

Since the resistance r_a of the ammeter and other connections has been included in the value of r', this resistance must now be subtracted from the computed result to obtain the resistance of the battery alone. Thus

$$r = r' - r_a$$

3-18. Slope of the Curve. By looking at the steepness of curves like those shown in Fig. 3-13 one can tell which batteries have the greater internal resistance. The very steep curves correspond to batteries of large internal resistance, and the low-resistance batteries show curves that are nearly horizontal. When such a curve has been obtained for a given battery, the internal resistance is given by the negative slope of the curve, Eq. (3-9).

3-19. Measurement of a Resistance Containing an EMF. The resistance of a circuit in which there is also an emf can be measured without the value of this emf being known. This problem might be the measurement of the internal resistance of a battery, as in the preceding sections, or it might be the determination of the resistance of a metallic circuit containing one or more emfs. If sufficient current can be drawn from the circuit, no other battery is required.

3-20. Useful Power from a Source. Electrical power is measured by the product EI, where I denotes the current and E the fall of potential over the circuit in which the power is being expended. The unit of power is the watt, one watt being the product of one volt by one ampere.

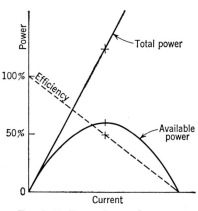

FIG. 3-14. Power from a battery.

When a battery is furnishing a current, the total power expended is supplied by the chemical reactions within the cell. Part of this power is expended in the external circuit, where it may be used in running motors or doing other useful work. The remainder is spent within the cell and goes only to warm the contents of the battery. In some cases the greater part of the energy is thus wasted within the cell.

The object of this experiment is to measure the power in the external circuit when various currents are flowing. The cell is joined in series with an ammeter and a resistance which will carry the largest current that may be used, as shown in Fig. 3-12. A voltmeter measures the fall of potential. Probably there will be found a point beyond which the useful power decreases even though the current is made larger. Since the current is proportional to the amount of chemicals used up in the cell, it will be well to express the results as a function of the current. This is best done by means of a curve, using values of the current for abscissas and the corresponding values of power for the ordinates.

The curve showing the total power EI supplied by the cell can also be plotted. The ratio of the ordinates of these two curves, that is, the ratio of the useful power to the total power, gives the efficiency of the cell.

The efficiency can also be plotted as a curve on the same sheet with the other two curves.

Record the observations as follows:

USEFUL POWER FROM A . . . CELL

R $= \dfrac{E'}{I}$	Ammeter		$I,$ amperes	Voltmeter		$E',$ volts	W $= E'I,$ watts
	Zero	Reading		Zero	Reading		

3-21. Maximum Amount of Available Power.

The amount of available power corresponding to a given current OC is shown graphically in Fig. 3-15 by the area of the rectangle $OABC$, whose sides are corresponding values of E' and I. When I is small, the area of this rectangle is small, increasing as I increases. After the mid-point is reached the relative decrease of E' is greater than the relative increase in I, and the area of the rectangle decreases to zero. Plotting the areas of the rectangles ($= W$) against the corresponding values of the current gives the curve of available power (Fig. 3-14).

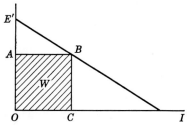

FIG. 3-15. Showing the amount of available power W corresponding to a given current ($= OC$).

At the point where the power W is a maximum, the curve between W and I becomes parallel to the I axis, and in the notation of the calculus,

$$\frac{dW}{dI} = 0$$

The relation between W and I is

$$W = E'I$$

where both E' and I are variables. Substituting for E' its value in terms of I from Eq. (3-5), Sec. 3-13, gives, with the ammeter removed,

$$W = EI - I^2r$$

and the derivative of this is

$$\frac{dW}{dI} = E - 2Ir$$

For a maximum value of W, this expression is equal to zero, which is the case when the current is

$$I = \frac{E}{2r}$$

When the current has this value, the battery is delivering more power than when the current is larger or smaller. Since for this current the resistance of the entire circuit is $2r$ and that of the battery is r, it follows that the resistance of the external circuit is also equal to r. Therefore the power *delivered* by the cell is a maximum when the external resistance is equal to the internal resistance. For power from a distant source delivered over a long line, all the resistance up to the point where the power is delivered should be counted as "internal" resistance.

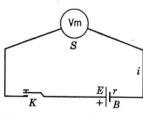

Fig. 3-16. A voltmeter does not measure the full emf of a battery.

3-22. Why a Voltmeter Does Not Measure the Full EMF of a Battery. When a cell B is joined to a voltmeter, the reading of the latter strictly does not give the emf of the cell. This discrepancy is small for cells of low internal resistance, but in many cases it cannot be neglected. If the resistance of the voltmeter is S and that of the cell is r, then by Ohm's law the small current through the cell and voltmeter is

$$i = \frac{E}{S + r}$$

From this the emf of the cell is

$$E = Si + ri$$

where Si is the fall of potential over the external circuit, in this case the voltmeter. Since a voltmeter measures only the fall of potential over its own resistance, the reading of the voltmeter is not E but Si, which is less than E by the amount ri. This term can be neglected when r is small, but not otherwise, unless i can be made very small.

In case r is known, a correction may be added to the voltmeter reading to obtain the value of E. If the above equation is written in the form

$$E = Si\left(1 + \frac{r}{S}\right) = E' + \frac{r}{S}E'$$

it is seen that the last term can be added to the voltmeter reading E' to obtain the full value of E.

3-23. How a Voltmeter Can Be Made to Read the Full EMF of a Battery.

In this method no current is taken from the battery, and therefore its full emf can be measured. The battery B is connected to a voltmeter as in Fig. 3-17, and then a second circuit is formed, consisting of the voltmeter, another battery Ba larger than B, and a variable resistance R.

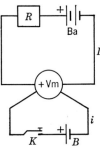

FIG. 3-17. To measure the emf of B.

Since there is no key in this second circuit, the voltmeter stands deflected and its reading V indicates the difference of potential between its terminals. When the key K is closed, the current through it may be either in one direction or in the opposite direction, or it may be zero, depending upon the value of V. If V is small, the current through K will be in the direction determined by the cell B, and the voltmeter will carry the combined current, $I + i$. If the potential difference V is large, it will cause a current to flow through the cell B in opposition to its emf E. In this case the main current I from the battery Ba divides at the voltmeter and flows through the two branches in parallel, just as though B were not present, except that the current in the lower branch is smaller than it would be with B absent.

If i denotes the value of this reverse current through B, the voltmeter current is $I - i$. The change in the voltmeter current from I to $I + i$, or to $I - i$, when the key is closed, will cause a change in the voltmeter reading, and no other ammeter is needed to tell the direction of the current through B.

Since the current through B can be in either direction, there is one point when this current will be zero. This will be when the tendency for the current to flow through B and K, due to the potential difference V across the voltmeter, is just balanced by the tendency for the current to flow in the other direction due to E. For this case of no change in the voltmeter deflection when K is opened or closed, there is no current through B, and $E = V$. The value of E can now be read from the voltmeter.

Measure in this way the emf of several cells and compare the results obtained in each case with the readings of the voltmeter when it is used alone.

The readings may be recorded as follows:

Name of cell	Voltmeter readings			E
	Zero	Alone	With aux. battery	

3-24. Measurement of Current by a Voltmeter and Shunt.

When a current I flows through a resistance R, the fall of potential over R is

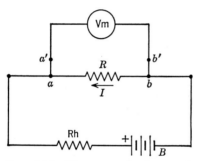

$E = IR$, which is in accordance with Ohm's law and the definition of the term "fall of potential." In Sec. 3-8 both E and I were measured and the value of R was then computed. When R is known, the experiment can be reversed and by measuring E the value of I can be computed. Thus a voltmeter may be used to measure currents in place of an ammeter. The arrangement is shown in Fig. 3-18.

Fig. 3-18. Measurement of current through Rh by a known resistance and a voltmeter.

Usually the resistance that would be suitable for this purpose is less than an ohm. Therefore any variation in the position of the voltmeter connections at a and b would cause considerable change in the resistance between these points. It is best to have these connections soldered fast and to let the voltmeter connections be made at the auxiliary points a' and b', where a little resistance, more or less, in the voltmeter circuit will be inappreciable.

Such shunts are often made having a resistance of 0.1, 0.01, 0.001 ohm or less. The current through the shunt is then 10, 100, 1,000 or more times the voltmeter reading. If the small current through the voltmeter is enough to be counted, it should be added to the current through the shunt to give the value of the total current that is measured by this arrangement.

This principle is also used in the construction of ammeters for measuring large currents. The greater part of the current is carried by a shunt of low resistance, while the delicate moving coil carries only a small current and thus in reality acts as a sensitive voltmeter. The numbers on the scale, however, instead of reading volts are made to give the corresponding values of the currents passing through the instrument.

3-25. Potential Divider. It is often necessary to use only a part of the given emf. This may be because a large emf is to be measured by a low-reading instrument, or it may be that a small fraction of a volt is required for use in some experiment. For this purpose, the source E is joined to a resistance AB that is large enough not to draw too much current from E or to become overheated itself. C is a sliding contact that can be moved from A to B, dividing the resistance in two parts, P and Q. The fall of potential between A and C is

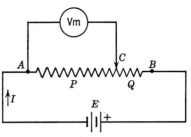

FIG. 3-19. Showing the use of a potential divider AB.

$$v = PI$$

and this value can be varied from zero to the full amount as desired by moving the sliding contact C.

A resistance AB that can be thus divided in various ratios is called a "potential divider." The emf E is divided into two parts, PI and QI, and by moving C, PI can be made any desired fraction of E.

The same effect can be obtained by using two resistance boxes for P and Q. By changing the ratio of the resistances in P and Q, the fall of potential across AC can be varied.

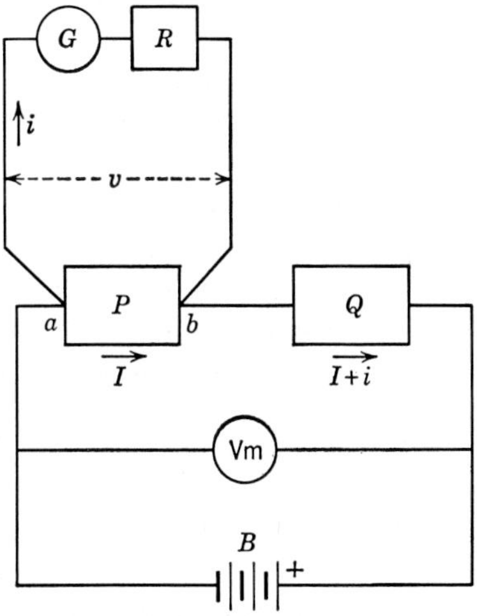

3-26. To Use a Potential Divider. Join the ends of a sliding rheostat to a constant emf. Connect a voltmeter to the sliding contact and to one end of the rheostat. Observe how the reading of the voltmeter varies as the sliding contact is moved from one end to the other. Note that moving the sliding contact does not affect the amount of current that is drawn from the battery, as would be the case when the rheostat is used as a variable resistance.

FIG. 3-20. A small voltage v is obtained from the potential divider PQ.

In the same way, the potential difference between A and C can be used wherever a fraction of the emf of E is desired.

3-27. To Compute the Ratio of a Potential Divider. The arrangement in Fig. 3-20 shows the potential divider PQ connected across the battery

B. The total fall of potential over $P + Q$ is V, and this is measured by the voltmeter Vm, but only the part PI is applied to the circuit of $G + R$.

Approximate Formula. When the current i through $G + R$ is negligibly small compared with the current I through the potential divider, the voltage v across P, which is applied to the circuit $G + R$, is readily computed by the relation

$$v = PI = P \frac{V}{P + Q} = \frac{P}{P + Q} V$$

This approximate value is often sufficiently accurate.

Exact Formula. In case the exact value of v is required, it is necessary to consider the small current i. If this is comparable to the current through P, the correction will be appreciable. The combined resistance of P and $(G + R)$ in parallel is

$$p = \frac{P(G + R)}{P + (G + R)}$$

and this is the effective resistance between a and b (Fig. 3-20).

The voltage that is applied to the circuit $G + R$ is equal to the fall of potential between a and b, which is

$$v = p(I + i) = p \frac{V}{p + Q} = \frac{p}{p + Q} V$$

where the value of p is computed by the formula given above.

3-28. Measurement of a High Resistance by a Voltmeter Alone. This method, a modification of the ammeter and voltmeter method for the measurement of moderate resistances, is based on the fact that a voltmeter is really a very sensitive ammeter. It can be used as an ammeter whenever the high resistance in series with the moving coil will not interfere with the desired measurement of current.

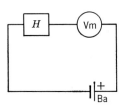

Fig. 3-21. To measure the high resistance H.

The voltmeter is joined in series with a battery and the high resistance H to be measured. In this way the voltmeter serves as an ammeter to measure the current through the circuit. The reading V of the voltmeter gives the fall of potential over its own resistance S or, in symbols,

$$V = Si$$

When the entire circuit is considered, the value of the current is given by the expression

$$i = \frac{E}{H + S}$$

Eliminating i from these two equations and solving for H gives

$$H = S\frac{E - V}{V}$$

The value of E, the emf of the battery, is easily measured by the same voltmeter by connecting it directly to the battery. A key arranged to short-circuit the high resistance will readily change the voltmeter from its position as an ammeter to that of a voltmeter.

When the emf of the source is too large to be measured by the voltmeter, or when for any other reason it cannot be determined, the resistance can still be measured. In the circuit shown in Fig. 3-21, the current is

$$i = \frac{V}{S} = \frac{E}{H + S}$$

Now add a large known resistance R in series with H and the voltmeter. Then the current through the circuit will be

$$i' = \frac{V'}{S} = \frac{E}{H + S + R}$$

Dividing one equation by the other gives

$$\frac{V}{V'} = \frac{H + S + R}{H + S}$$

and

$$H = \frac{V'(S + R) - VS}{V - V'}$$

The observations can be recorded as follows:

To Measure the Resistance of . . .

Vm zero = Vm resist. = . . . ohms

Name of object measured	V	E	H

3-29. To Make a High-reading Voltmeter from a Low-reading Instrument. Voltmeters are made in different ranges. Some will measure voltages over the range of 0 to 3 volts; others over the range 0 to 15 volts or 0 to 150 volts, etc. In each of these instruments the moving system is the same and requires the same amount of current to operate it. The principal difference lies in the resistance of the voltmeter. If the resistance is 300 ohms, an emf of 3 volts will give a current of 0.01 amp through the voltmeter, and this is sufficient to deflect the movable pointer to the upper end of the scale. The range of this voltmeter would be 0 to 3 volts.

If the same instrument had a resistance of 1,500 ohms, it would require 15 volts to give a current of 0.01 amp and deflect the pointer to the high end of the scale. The range would be 0 to 15 volts, and the instrument would be called a "15-volt voltmeter." If the resistance were increased to 15,000 ohms, the voltmeter could measure up to 150 volts.

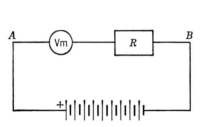

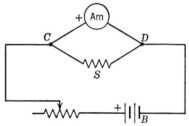

FIG. 3-22. A large resistance R in series with a low-reading voltmeter Vm makes the combination AB into a high-reading voltmeter.

FIG. 3-23. A small resistance S connected in parallel with a low-reading ammeter Am makes the combination CD into a high-reading ammeter.

3-30. To Make a High-reading Ammeter from a Low-reading Instrument. An ammeter measures the current that passes through it. Such an instrument is placed in series with the circuit in which the value of the current is desired. The moving system and general appearance of an ammeter are like those of a voltmeter, and when the current to be measured is small, it passes directly through the moving coil and causes the deflection of the pointer over the scale. Adding more resistance in series will reduce the value of the current, but the same value of the current will deflect the pointer to the same place on the scale of the ammeter.

When it is desired to measure currents that are larger than are indicated by the upper end of the ammeter scale, it is necessary to arrange the circuit so that the ammeter itself measures only a certain fraction of the total current. Thus if a shunt of 0.05 ohm is placed in parallel with an ammeter whose resistance is 0.05 ohm, the current will divide, half passing through the ammeter and the other half through the shunt. The current measured by the ammeter should be multiplied by 2 to give the value of the total current. This combination of an ammeter with a shunt of equal resistance makes an effective ammeter of twice the range of the ammeter alone. If a shunt of smaller resistance is used, a larger part of the current will pass through the shunt, and the range for the total current measured by the ammeter and shunt is increased. Ammeters for measuring large currents are thus provided with a built-in and permanent shunt of low resistance. The uncertain resistance of the contact connections of the shunt and ammeter will introduce uncertainty in the fraction of the current that will pass through the ammeter branch of the circuit unless these connections are soldered permanently in a fixed position.

3-31. Time Test of a Cell. The time test of a cell is designed to show how well the cell can maintain a current, how effective is the action of the depolarizer, and the rapidity and extent of the recovery of the cell when the current ceases. Such a test usually continues for an hour, and the results may best be shown graphically by means of curves. These curves should show the value of the emf of the cell at each instant during the test, the available emf, the current, and the internal resistance of the cell. After the current is maintained for an hour, the circuit is opened and the cell is allowed to recover. The curves should show the rate and extent of this recovery.

Of course it is evident that such a continued test may prove rather severe for cells intended for only a few minutes' use at one time. On the other hand, cells that make a good showing under the test might not be the best for long-continued intermittent service. Nevertheless, such a test gives the most information regarding the behavior of the cell that can be obtained in the short space of 2 hr.

The setup for making a time test is the same as that for measuring the internal resistance of the cell. The battery circuit remains closed all the time, however, except when it is opened for an instant to measure the emf of the cell. With the proper preparation beforehand it is not difficult to observe all the necessary data and record them in a neat and convenient form.

Readings should be taken once a minute. Practice in doing this should be obtained by measuring the emf of the cell before any current is drawn from it. When ready to commence the test proper, close the circuit through 5 or 10 ohms, or about as much resistance as the internal resistance of the cell, and as soon thereafter as possible take the first reading for the available emf. One minute later, open the circuit just long enough to take a reading of the emf of the cell. Thus every 2 min the available emf is recorded and on the intermediate minutes the value of the total emf is measured.

From these data two curves are plotted, one showing the variation in the emf of the cell during the test and the other showing the same with respect to the available emf. The internal resistance is computed at 5-min intervals, the values of E and E' being taken from the curves. The current is measured by the ammeter, or it may be computed from the values of R and the available emf.

After the first hour of the test, the battery key is changed so as to keep the circuit open, except for an instant each minute when it is closed long enough to read the voltmeter. Since the voltmeter draws some current from the cell, and as it is sometimes difficult to obtain simultaneous readings of the ammeter and voltmeter at the instant desired, more accurate results are possible when the measurements are made by the condenser and ballistic-galvanometer methods described in a later chapter. The keys can be worked by hand, and if proper care is exercised, good results may be expected. Better results may be obtained by using a special battery-testing key or a pendulum apparatus that will close and open the keys in precisely the same manner each time.

During the second hour the battery will recover more or less completely from the effects of polarization, and at the end of the 2-hr test the value of E should be about the same as at the beginning. The recovery curve may be plotted backward across the sheet containing the other curves, thus showing very clearly the extent of the recovery.

The observations may be recorded as below.

Time Test of a . . . Cell

Time of day		E	E'	I	r
Hour	Minute				

REFERENCES

Stout, Melville B.: "Basic Electrical Measurements," Prentice-Hall, Inc., Englewood Cliffs, N.J., 1950.

Michels, Walter C.: "Electrical Measurements and Their Applications," D. Van Nostrand Company, Inc., Princeton, N.J., 1957.

PROBLEMS

3-1. An ammeter has a coil resistance of 10 ohms and gives a full-scale deflection when a current of 1 ma passes through the coil. How can this meter be converted into a 10-ma meter?

3-2. A 0- to 10-volt voltmeter is made from an ammeter which has a coil resistance of 20 ohms and a sensitivity of 0 to 1 ma by placing a resistance of 9,980 ohms in series with the meter coil. Compute the value of the resistance which must be placed in parallel with the 20-ohm coil to convert the 0- to 10-volt voltmeter into one which reads 0 to 50 volts. What is the total resistance between the terminals of the meter both as a 0- to 10- and a 0- to 50-volt voltmeter?

3-3. With the meter movement described in Prob. 3-2, compute the resistance which must be placed in series with the coil to convert it to a 0- to 50-volt voltmeter. What is the resistance between the terminals? Discuss the advantages and disadvantages of these two methods for producing a 0- to 50-volt voltmeter.

3-4. A shunt is rated as 50 mv at 300 amp. This shunt is used with a meter which has a coil resistance of 25 ohms and which gives a full-scale deflection with 5 ma. What is the range of the shunted meter? What is the resistance of the shunt?

3-5. The shunt of Prob. 3-4 is used with a meter which has a coil resistance of 50 ohms and a range of 0 to 1 ma. What is the range of the shunted meter?

3-6. An ammeter has a range of 0 to 1 amp. Its coil is replaced with a coil which has 50 per cent more turns and an area 5 per cent greater. The torsional constant of the suspension system is 30 per cent less. What is the range of the rebuilt meter?

3-7. A potential divider is constructed from a resistance of 1 ± 0.01 ohm in series with a resistance of 50 ± 1 ohms. The voltage across the combination is measured as 2.10 ± 0.05 volts. What is the voltage across the 1-ohm resistance?

3-8. A voltmeter of 20 ohms resistance is used to measure the voltage across the 1-ohm resistance. What is the measured value of the voltage?

3-9. When the terminal voltage of an old cell is measured with a voltmeter which has a resistance of 100 ohms, a reading of 1.23 volts is obtained. The measurement is repeated with a 30-ohm voltmeter, and a reading of 0.63 volt is obtained. What is the emf of the cell?

3-10. A cell has an emf of 20 volts and an internal resistance of 5 ohms. Calculate the power delivered to each of the following resistances when connected across the cell: 1, 2, 3, 4, 5. 6, 8, 10, and 20 ohms.

CHAPTER 4

CIRCUIT THEOREMS AND THEIR APPLICATIONS
TO DIRECT CURRENT CIRCUITS

4-1. Introduction. When making measurements one often encounters circuits which contain several sources of voltage and a number of resistors. If such circuits are of the simple series type, they can be reduced to an equivalent circuit of one voltage source in series with a single resistance. The voltage source in the equivalent circuit is the algebraic sum of all the original voltage sources, and the equivalent resistance is equal to the sum of all the original resistances. After such a reduction the circuit can be solved by means of Ohm's law. For a parallel arrangement of resistances or a parallel-series combination, a similar procedure is applicable. If the configuration of the elements becomes more complex, such a simple reduction is no longer completely adequate. It is then necessary to resort to a more elaborate method for solving the circuit. In the following paragraphs we shall consider several of the more important circuit theorems in some detail. Although the procedures are presently applied to d-c circuits, they are equally applicable to a-c circuits.

4-2. Kirchhoff's Laws. Kirchhoff's first law can be stated as follows: At any branch point in a circuit, the sum of all the currents flowing toward the point is equal to the sum of the currents leaving the point.

A branch point in a circuit is a junction at which several wires meet. For the branch point shown in Fig. 4-1, the first law can be written in the form of an equation:

$$I_1 + I_2 + I_3 = I_4 + I_5$$

Basically the law can be reduced to a statement that, in effect, electrical charge acts like an incompressible fluid and therefore does not pile up at a point in the circuit. Since a steady current can be expressed as q/t, it follows that the currents entering and leaving the branch point are equal.

Kirchhoff's second law can be stated as follows: In going around any *complete* electrical circuit, noting the voltage change across each element traversed (in both magnitude and sign), the algebraic sum of these voltage changes is zero.

The second law has its foundation in the fact that potential, being work per unit charge, is a scalar quantity and therefore depends only upon the starting and ending point, not upon the path which is followed. Since the starting and ending points are one and the same if a complete circuit is traversed, the sum of the voltage increases must just equal the sum of the voltage drops.

In writing equations derived from the second law, it is of great importance to adopt a consistent convention concerning the signs of the voltage changes which are encountered. In d-c work, two main types of voltage changes are encountered. A voltage change will occur across a resistance when a current flows through the resistance; in addition, voltage changes

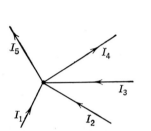

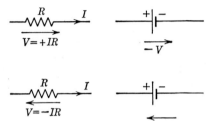

FIG. 4-1. Branch point, illustrating Kirchhoff's first law.

FIG. 4-2. Conventions for the voltage changes in applying Kirchhoff's second law.

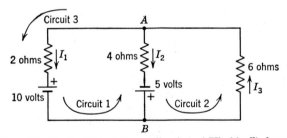

FIG. 4-3. Illustration of the application of Kirchhoff's laws.

may be produced in the circuit as a result of chemical, mechanical or thermal changes in certain elements.

One system which is convenient is indicated in Fig. 4-2. When traversing a resistance in the direction of flow of current (electron), the voltage change is $+IR$; in going across the resistance in the direction opposite the current, the voltage change is $-IR$. If a source of emf such as a battery is encountered, the voltage change will be $-V$ when going from the positive to the negative terminal and $+V$ when going in the opposite direction. (It is assumed that the internal resistance of the source is zero. If the internal resistance of the source is not zero, it can be treated as an ordinary resistance in series with the source.)

Kirchhoff's laws can be applied to solve the circuit shown in Fig. 4-3. In this circuit, the components (sources of voltage and resistances) are given and it is necessary to find the current flowing through each. It is customary to indicate the direction of the current through each element by means of an arrow. This direction can be chosen arbitrarily at this point. Although some of these directions may, indeed, be chosen incorrectly, the work will not be impeded, since the answer, when obtained, will give both the magnitude and direction of the currents. A positive value for a current indicates that the original choice for the direction of the current was correct. A negative sign will indicate that the current is actually flowing in the direction opposite that which was originally chosen.

Since the circuit under consideration has three unknown quantities, it is clear that three independent equations are required to effect a solution. The application of the first law at point A yields the first of these equations:

$$I_3 = I_1 + I_2 \tag{4-1}$$

The corresponding equation for branch point B is identical and would yield no additional information.

The second law may be applied to circuit 1. Starting at the point A and moving in a counterclockwise direction,

$$2I_1 - 10 - 5 - 4I_2 = 0 \tag{4-2}$$

Circuit 2 when traversed in a counterclockwise direction yields a similar equation:

$$4I_2 + 5 + 6I_3 = 0 \tag{4-3}$$

Substituting the value of I_3 from Eq. (4-1) into Eq. (4-3) yields

$$10I_2 + 6I_1 + 5 = 0$$

Multiplying Eq. (4-2) by -3 gives $12I_2 - 6I_1 + 45 = 0$. The latter two equations are added to give $I_2 = -\frac{50}{22}$ amp. The negative sign indicates that the current through the 4-ohm resistor actually flows in the direction opposite that indicated by the arrow. The value obtained for I_2 is used to complete the solution: $I_1 = \frac{65}{22}$ amp and $I_3 = \frac{15}{22}$ amp.

It would have been possible to use the third circuit to obtain one of the equations of the second law. It should be stressed that this third circuit would be used instead of one of the other two, not in addition to them, since in writing any two of these equations, one has used each of the circuit elements at least once. The third equation would therefore contain no new information and, in fact, could be derived by combining the first two circuit equations.

4-3. Superposition Theorem. For a circuit which contains only linear elements[1], the resulting current through each element is made up of the

---

[1] A linear element is one for which the current is directly proportional to the impressed voltage.

currents which would be produced by each source of voltage acting on the circuit separately. That is, the currents from each source are superimposed upon each other to yield the total current through any circuit element.

This theorem reduces the problem of calculating the currents in a circuit to a series of problems, each of which can be solved independently by the application of Ohm's law. There will be one problem of the latter type for each source of voltage present in the original circuit.

This principle can be illustrated by means of the circuit shown in Fig. 4-3. We shall first find the effect which the 10-volt battery would have upon the circuit. The circuit can be redrawn (Fig. 4-4) with all other voltage sources removed. (If a source has an internal resistance, this must, of course, be retained.)

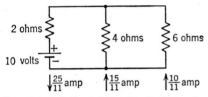

FIG. 4-4. Solution of the circuit shown in in Fig. 4-3 by means of the superposition theorem. Effect of the 10-volt battery.

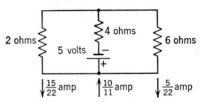

FIG. 4-5. Effect of the 5-volt battery on the circuit.

The procedure is then repeated for each of the other sources of voltage (Fig. 4-5). It is seen that the 10-volt battery produces a current of $25/11$ amp downward through the left branch while the 5-volt cell produces a current of $15/22$ amp downward through this branch. In the original circuit both sources are acting simultaneously to produce a downward current of $65/22$ amp. Similarly the current through the middle branch is $50/22$ amp upward while the resultant current through the right branch is $15/22$ amp upward.

From the standpoint of the superposition theorem, zero current in a given branch would arise from a system which produced two currents of equal magnitude but in opposite directions through the branch.

4-4. Thévenin's Theorem. In its most elementary form, Thévenin's theorem can be stated as follows: Any complex circuit can be broken into two parts; one of these parts can be replaced by an equivalent circuit which consists of one voltage source in series with a single resistance. The equivalent voltage source has a magnitude equal to the potential difference across the open terminals of the part of the circuit to be replaced. The equivalent resistance is the resistance between these open terminals when all voltage sources have been replaced by their internal resistances. (It frequently happens that the internal resistance of a battery is negligibly small. In this case the battery is replaced by a resistanceless wire.)

The use of this theorem is illustrated with the circuit indicated in Fig. 4-6. The circuit is broken at the points A and B. The 6-ohm resistance is retained, while the more complex part of the circuit to the left of A-B will be replaced by a single resistance and a single voltage source.

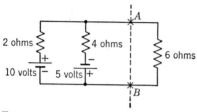

FIG. 4-6. Application of Thévenin's theorem. The portion to the left of A-B is replaced by an equivalent resistance in series with an equivalent voltage.

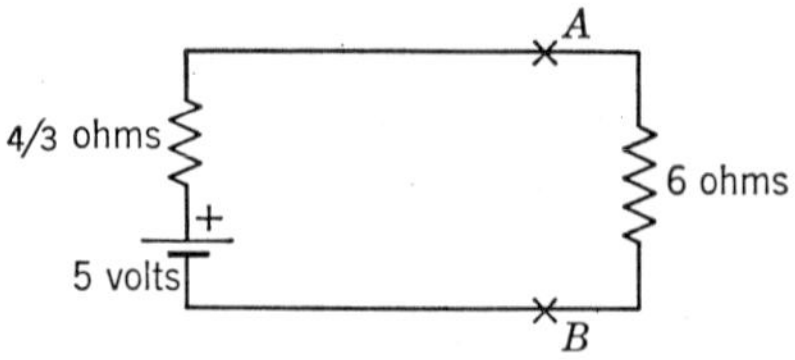

FIG. 4-7. Thévenin's equivalent circuit for Fig. 4-6.

Considering only the part to the left of A-B, an electron current will flow counterclockwise with a magnitude

$$I = \frac{10 + 5}{4 + 2} = \frac{15}{6} \text{ amp}$$

Starting at point A we shall add up all the voltage changes until we reach point B. This yields $V_{\text{equivalent}} = (-15/6 \times 4) + 5$, or -5 volts.

The batteries are now removed and replaced by wires (we have assumed zero internal resistances). The resistance to the left of A-B consists of a 2- and a 4-ohm resistance in parallel. Thus $R_{\text{equivalent}} = \frac{4}{3}$ ohms.

The equivalent circuit is shown in Fig. 4-7. This immediately yields the current through the 6-ohm resistance, $\frac{15}{22}$ amp upward. We have thus obtained one piece of information about the original circuit, the current through the 6-ohm resistance. With this value we can now return to the original circuit to find the other two remaining currents.

As a second illustration of Thévenin's theorem, consider the circuit shown in Fig. 4-8. This is basically the circuit of an unbalanced Wheatstone bridge in which R_g represents the resistance of the galvanometer. Again a break is made at A-B. R_g is retained, and the remaining portion of the circuit is replaced by a single equivalent battery and an equivalent resistance according to the procedures already stated. In Fig. 4-9 it is seen that the current in the upper branch is $I_U = \frac{1}{25}$ amp while the current in the lower branch is $I_L = \frac{1}{15}$ amp. The potential difference between A and B is $V_{\text{equivalent}} = -\frac{1}{25} \times 20 + \frac{1}{15} \times 10$, or $-\frac{2}{15}$ volt.

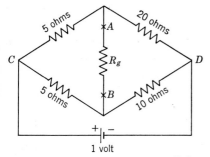

Fig. 4-8. Thévenin's theorem applied to an unbalanced Wheatstone bridge.

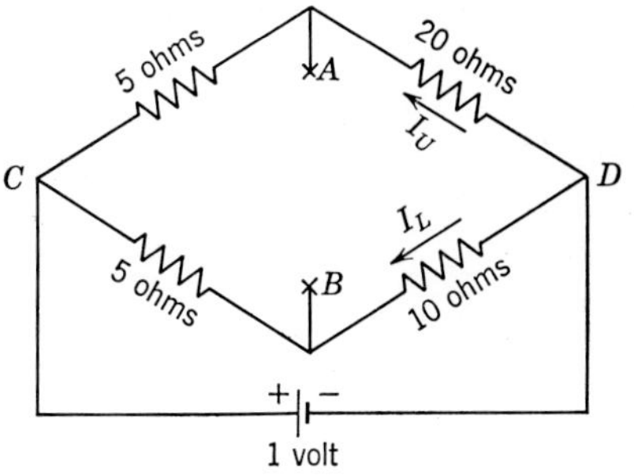

Fig. 4-9. Portion of a circuit which will be replaced by an equivalent circuit.

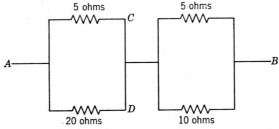

Fig. 4-10. Equivalent resistance is the resistance between A and B.

To find the equivalent resistance between A and B with the battery shorted out, the circuit is redrawn in Fig. 4-10.

$$R_{A\text{-}B} = R_{\text{equivalent}} = {}^{110}\!/_{15} \text{ ohms}$$

The final circuit (Fig. 4-11) immediately yields the current through the galvanometer.

$$I_g = \frac{{}^{2}\!/_{15}}{{}^{110}\!/_{15} + R_g}$$

4-5. Proof of Thévenin's Theorem. Thévenin's theorem can be shown to be a natural consequence of the superposition theorem. Figure 4-12 is a general circuit containing a number of sources of voltage and resistances. The potential difference between the points A and B is $V_{A\text{-}B}$. At this point let us add an additional branch between A and B consisting of a pure resistance R, as shown in Fig. 4-13. This will alter the conditions which had existed in the previous circuit (Fig. 4-12). The problem is to find the current which flows through the resistance R.

If a further change is made in the circuit, the current through R can be reduced to zero. This is accomplished by inserting a battery in series with R of magnitude $V_{A\text{-}B}$ and in such a direction as to oppose the voltage which already exists between the points A and B. According to the superposition theorem, this zero of current can be considered to be made up of two equal and opposite currents, one supplied by the circuit in Fig. 4-12, the second by the battery V_{A-B} which has been inserted in series with R. The problem has now been reduced to that of finding the current

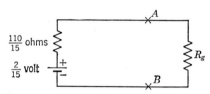

Fig. 4-11. Thévenin's equivalent circuit for the unbalanced Wheatstone bridge.

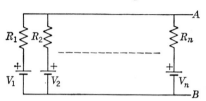

Fig. 4-12. Circuit for proof of Thévenin's theorem.

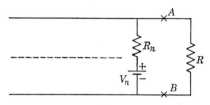

Fig. 4-13. Thévenin's theorem is a natural outgrowth of the superposition theorem.

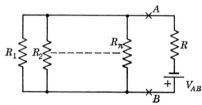

Fig. 4-14. The current supplied by V_{A-B} is equal and opposite to that supplied by the circuit in Fig. 4-12.

furnished by the battery V_{A-B}. This will be equal in magnitude but in the opposite direction to that supplied by the circuit in Fig. 4-12. The diagram used is shown in Fig. 4-14.

$$I_R = \frac{V_{A-B}}{R + R_{\text{equivalent}}}$$

$R_{\text{equivalent}}$ is just the equivalent resistance between A and B of the original circuit with all the batteries shorted out.

4-6. Δ-Y Transformation. Strictly, the Δ-Y transformation is not a circuit theorem at all but merely a relationship between two circuits of the types shown in Fig. 4-15 such that one circuit is made to be equivalent to the other just as the reciprocal law allows us to replace several parallel resistances by a single equivalent resistance.

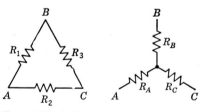

Fig. 4-15. A delta, Δ, and a wye, Y, circuit.

The transformation can be arrived at in the following manner:

In order that the Y be equivalent to the Δ, they must have a mutual relationship such that nothing in the circuit external to them will be influenced when one is substituted for the other. For example, the current flowing into the Δ at point A must be equal to the current flowing into the Y at point A, etc.

In the Δ circuit, current may flow from A to B by means of two paths. First, it may flow directly from A to B through the resistor R_1. A second parallel path is provided through R_2 and R_3 in series with each other.

In the Y circuit current can flow from A to B only through the series combination of R_A and R_B. Since the effective resistance between A and B for the Δ must equal the resistance between A and B for the Y, the following equation can be written:

$$\frac{1}{R_{A\ to\ B}} = \frac{1}{R_A + R_B} = \frac{1}{R_1} + \frac{1}{R_2 + R_3}$$

Similarly for the other points:

$$\frac{1}{R_{A\ to\ C}} = \frac{1}{R_A + R_C} = \frac{1}{R_2} + \frac{1}{R_1 + R_3}$$

$$\frac{1}{R_{B\ to\ C}} = \frac{1}{R_B + R_C} = \frac{1}{R_3} + \frac{1}{R_1 + R_2}$$

With these three equations, it is now possible to obtain R_A, R_B, and R_C in terms of R_1, R_2, and R_3. These relationships are

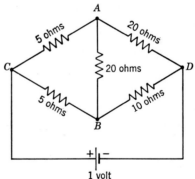

FIG. 4-16. Unbalanced Wheatstone bridge.

$$R_A = \frac{R_1 R_2}{R_1 + R_2 + R_3}$$

$$R_B = \frac{R_1 R_3}{R_1 + R_2 + R_3}$$

$$R_C = \frac{R_2 R_3}{R_1 + R_2 + R_3}$$

The connecting equations are easily remembered if it is noted that the resistor attached to a given point in the Y circuit is equal to the product of the two resistors attached to the corresponding point in the Δ circuit divided by the sum of all three resistances in the Δ circuit.

The application of this transformation can be illustrated by means of the unbalanced Wheatstone bridge (Fig. 4-16). The delta to the right of the circuit ABD will be replaced by an equivalent wye circuit as shown in Fig. 4-17. Ohm's law can now be applied to find the current in the upper branch as well as the lower branch. $I_{upper} = \frac{9}{205}$ amp, while

$$I_{lower} = \frac{13}{205} \text{ amp}$$

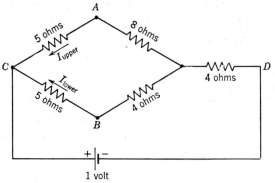

FIG. 4-17. The delta has been replaced by a wye.

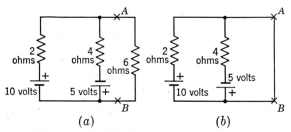

(a)   (b)

FIG. 4-18. Application of Norton's theorem.

The potential difference between A and B is

$$\tfrac{9}{205} \times 5 - \tfrac{13}{205} \times 5 = -\tfrac{4}{41} \text{ volt}$$

Returning to the original circuit (Fig. 4-16) the current flowing through the galvanometer from B to A is $\tfrac{4}{41} \div 20 = \tfrac{1}{205}$ amp. This may be compared with the method of Thévenin's theorem.

4-7. Norton's Theorem. Norton's theorem is similar in many respects to that of Thévenin. It again allows a portion of a complex circuit to be replaced by an equivalent circuit. The equivalent in this case consists of a current generator shunted by a resistance. The current supplied by the generator is equal to the current which would flow between the terminals of the circuit to be replaced if these terminals were shorted out. The equivalent resistance is the resistance between the open termi-

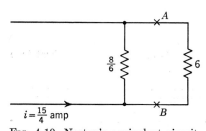

FIG. 4-19. Norton's equivalent circuit. A portion of the original circuit is replaced by a current generator shunted by a resistance.

nals of the circuit to be replaced and as such has the same value as the equivalent resistance in Thévenin's theorem.

Norton's theorem is applied to the circuit shown in Fig. 4-18. The 6-ohm resistance is removed, and a wire is connected between the points A and B. The current through the wire is $15\frac{3}{4}$ amp upward. The resistance between the open terminals A-B is $\frac{8}{6}$ ohms. The equivalent circuit is indicated in Fig. 4-19. The current through the 6-ohm resistance is $15\frac{3}{22}$ amp upward.

4-8. Reciprocity Theorem. Consider the influence of only one source of voltage upon a circuit. (There may, indeed, be other sources of voltage which furnish current, but by means of the superposition theorem it is possible to investigate the influence of each separately.) According to the reciprocity theorem, if this source when at position X produces a certain current at position Y, the same amount of current will flow at point X if the source is placed at point Y.

PROBLEMS

4-1. Calculate the current through each resistance by means of Kirchhoff's laws.

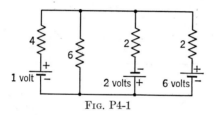

FIG. P4-1

4-2. Find the current flowing through each resistance by means of the superposition theorem.

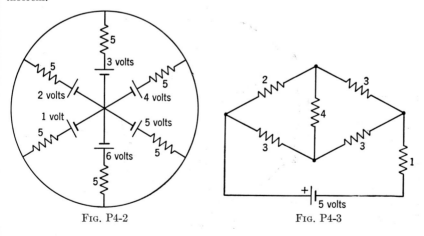

FIG. P4-2   FIG. P4-3

4-3. Compute the current through each resistance by the application of Thévenin's theorem.

4-4. Solve the circuit given in Prob. 4-3 by means of the Δ-Y transformation.

4-5. What is the effective resistance between the points A and B?

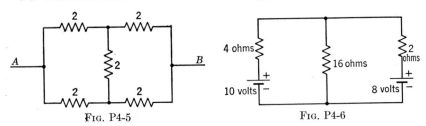

FIG. P4-5   FIG. P4-6

4-6. Apply Norton's theorem to find the current through the 16-ohm resistance. What current flows through the 2-ohm resistance?

4-7. Express each resistance of the Δ circuit in terms of the three resistances of the Y circuit.

4-8. Apply Thévenin's theorem by breaking the circuit at the points X-Y to find the current through the 20-ohm resistance.

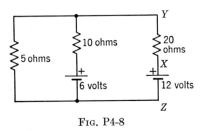

FIG. P4-8

4-9. Using the circuit of Prob. 4-8, apply Thévenin's theorem to find the current through the 20-ohm resistance.   Break the circuit at Y-Z.

4-10. Find the effective resistance between A-B, A-C, and B-C.

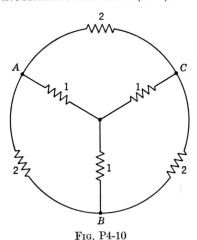

FIG. P4-10

CHAPTER 5

THE CURRENT GALVANOMETER

5-1. Introduction. The principles underlying the moving-coil ammeter have been discussed in Chap. 3. It was shown that a rectangular coil, when placed in a radial magnetic field, could be made to rotate through an angle proportional to the current passing through the coil. In its essential features, a current galvanometer is very similar to a portable ammeter. The chief difference arises in the nature of the suspension which holds the coil in the magnetic field and supplies the countertorque which tends to return the coil to its normal position. In the portable ammeter, the coil is held by a jeweled bearing arrangement and the restoring torque is supplied by a spiral spring attached to the coil. This leads to a very stable and rugged system, but it does so at the sacrifice of sensitivity. The galvanometer coil is suspended by a fine conducting fiber, usually made of gold. This not only holds the coil in place within the magnetic field but also supplies the countertorque which opposes the current-produced torque in the coil.

5-2. Physical Factors Which Influence the Sensitivity of a Galvanometer. The current passing through the coil of an ammeter or a galvanometer (Chap. 3) can be written in terms of the various physical parameters $I = \tau\theta/KBNA$. τ is the torque required to turn the coil through a unit angle, θ is the angular deflection of the coil, K is a constant which depends upon the units used, N is the number of turns on the coil, and A is the effective area of the coil. B is the magnetic field strength. For a suspension of circular cross section $\tau \sim \mu R^4/L$. μ is the torsional modulus of the material from which the suspension is made, R is the radius of the suspension, and L is its length. The angular deflection of an ammeter is usually measured by connecting a needle directly to the coil and observing the position of this needle or pointer on a scale attached to the meter. The deflection of a galvanometer is observed by attaching a small mirror on the coil and observing the reflection of a fixed scale through a telescope. If the scale is at a constant distance l from the mirror, the angular displacement expressed in radians is $\theta = d/2l$. d is the deflection on the scale measured in the same units as l.

It is now possible to express the ratio $I/d \sim \mu R^4/BANlL$. A very sensitive galvanometer will have a low ratio for I/d; that is, a small current will produce a relatively large deflection. Thus to build a very sensitive galvanometer one would want a suspension material such as gold which has a low torsional modulus. The suspension should be long, with a small radius. The magnetic field should be strong, and the coil should have many turns and a large cross section. It is probably clear that not all these quantities are independent of one another as, for example, the radius of the suspension and the coil. In making the number of turns and the area of the coil large one necessarily increases the weight of the coil which must be supported by the suspension, thereby limiting the minimum radius of the suspension. In addition to such obvious limitations, other factors such as the mechanical stability of the system, the period of oscillation, and coil resistance must be taken into account in designing a galvanometer.

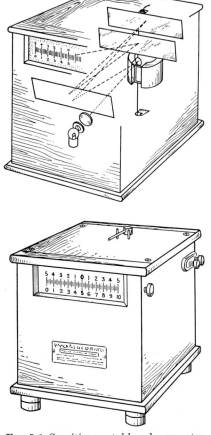

Galvanometers of rather high sensitivity and good mechanical stability are available (Fig. 5-1). Stability is obtained by using a short suspension system under tension. This by itself would tend to give low sensitivity. The higher sensitivity is obtained by greatly increasing the light path l by a multiple reflection system.

5-3. Damping of a Galvanometer. When a galvanometer coil is deflected through an angle θ, an amount of energy equal to $\int_0^\theta \tau\theta \, d\theta$ is stored in the system in the form of mechanical potential energy and resides in the twisted suspension.

Fig. 5-1. Sensitive portable galvanometer. (*Courtesy of the Rubicon Company.*)

When the current through the galvanometer is cut off, the coil will move toward its equilibrium position, and if no damping is present, it will reach the equilibrium position with a rotational kinetic energy of $\frac{1}{2}J\omega_{max}^2 = \frac{1}{2}\tau\theta_{max}^2$. J is the moment of inertia of the system, and ω_{max} its angular velocity. That is, all the potential energy originally stored in the twisted suspension will have been converted to rotational kinetic energy. This obviously leads to a condition which will cause the coil to oscillate in undamped simple harmonic motion. If one considers the friction present (in the suspension and also the friction between the moving coil and the air), the coil will oscillate with an amplitude which decreases with time. Under such conditions it may require many periods of oscillation before all the energy is dissipated and the coil again comes to rest.

It is possible to increase the damping effects on a galvanometer and thereby bring it to its initial position without oscillating. This involves the expenditure of the potential energy by electrical means rather than by mechanical friction. Commonly a single closed loop of heavy wire is added to the moving coil. When the coil moves in the magnetic field, a voltage E is generated in this loop and energy is dissipated at the rate of E^2/R joules/sec. By making the loop of the proper resistance R, it is possible to have the coil return to its zero position in a minimum of time without oscillating. The galvanometer would then be said to be critically damped.

An alternate method for producing critical damping can also be used. This is done by shunting the galvanometer coil with a certain resistance, its external critical damping resistance (CDRX). When the coil starts swinging, it acts as a generator and dissipates energy in the closed circuit consisting of the coil in series with the external critical damping resistance.

Under some circumstances it may be undesirable to use the latter method, since the presence of the damping resistance across the meter coil will make the system less sensitive. This may be particularly important with a galvanometer having a high coil resistance. The loop method, on the other hand, does not influence the sensitivity of the galvanometer; however, it does increase the moment of inertia and the period of the coil.

5-4. Figure of Merit. (*a*) *Direct Deflection.* Most galvanometers are so constructed that, for small angles at least, the deflection is directly proportional to the current. That is,

$$I = Fd$$

The factor F is called the "figure of merit" of the galvanometer, and it is defined as the *current per scale division* (1 mm) that will deflect the galvanometer. The figure of merit of most galvanometers is smaller than one hundred-millionth of an ampere per millimeter.

Inasmuch as the deflection will vary with the distance of the scale from the mirror, this distance should be recorded with the other observations. The standard value of F is computed for a distance of 1 m.

In order to determine the figure of merit, it is necessary to send a small known current through the galvanometer and observe the *steady* deflection it produces. The method can be understood by reference to Fig. 5-2. The galvanometer is joined in series with a battery, a large resistance, and a key. When the key is closed, the current through the circuit, and therefore through the galvanometer, is, by Ohm's law,

$$i = \frac{E}{g + r + R}$$

where E denotes the emf of the battery and g, r, R, the resistances of the galvanometer, battery, and R, respectively. If this current produces a steady deflection of d scale divisions (millimeters), the figure of merit is $F = i/d$.

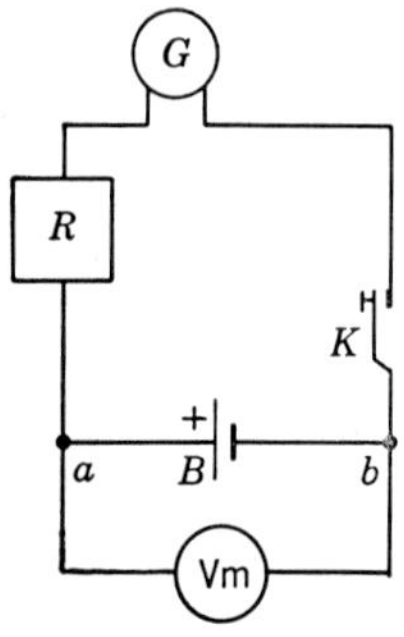

By using a voltmeter to measure directly the fall of potential between a and b, the current i will be given by

$$i = \frac{V}{R + g}$$

where V is the voltmeter reading. Then

$$F = \frac{V}{(R + g)d}$$

FIG. 5-2. To determine the figure of merit of G.

Thus, by the use of the voltmeter, the somewhat uncertain emf of the cell is replaced by the definite voltmeter reading, and the unknown resistance of the cell does not appear in the equations. Of course V and d must be simultaneous values, and it is understood that d is the *steady* deflection produced by the steady current i.

5-5. Figure of Merit. (b) *Potential-divider Method.* With a sensitive galvanometer it is usually not possible to make R large enough to use the simple method. It is then most convenient to use a value for E that is only a small fraction of the emf of the cell. This can be done by the potential-divider method shown in Fig. 5-3. The fall of potential between a and b is now PI instead of E, where each factor P and I can be made as small as necessary. For most galvanometers it is convenient to make $P + Q = 1{,}000$ ohms, and R about 100,000 ohms.

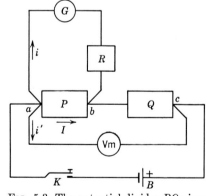

FIG. 5-3. The potential divider PQ gives across ab a small fraction of the emf of the battery B.

The current from the battery divides at a, one part, i, going through the galvanometer, another part, I, through P, and the third part flowing through the voltmeter. The fall of potential from a to b across P is the same as through R and the galvanometer, or

$$PI = (R + g)i \tag{5-1}$$

where g is the resistance of the galvanometer.

The current through Q is $I + i$, and the fall of potential from a to c, which is measured by the voltmeter, is

$$PI + Q(I + i) = V \tag{5-2}$$

Eliminating I between Eqs. (5-1) and (5-2) and solving for i give

$$i = \frac{PV}{(Q + P)(R + g) + PQ}$$

The figure of merit is, then,

$$F = \frac{PV}{(P + Q)(R + g) + PQ} \frac{1}{d}$$

It is evident that the computed value of F cannot be any more certain than are the measured values of V and of d.

It is always a wise precaution not to connect the battery into the circuit until after the rest of the setup is completed and has been carefully examined to make sure that no unintended connections have been made.

At the start, P should be set very much smaller than Q, perhaps $P = 1$, and $Q = 999$, so that the first deflection of the galvanometer will not be too large. When this has been tried, P can be increased enough to give a deflection of 100 or 200 mm. The observations can be recorded as follows:

To Determine the Figure of Merit of Galvanometer No. . . . .

Zero	Reading	Deflection	R	V	P	Q	F

When scale to mirror is 1 m, $F =$

In case the values of F are nearly the same for large and small deflections they can all be averaged to give a mean value of F for the galvanometer.

But, when the values of F show a continuous variation as the deflections become larger, it is better to show these values by a curve as in Fig. 5-4. Then the proper value of the figure of merit for any given deflection is read directly from the curve.

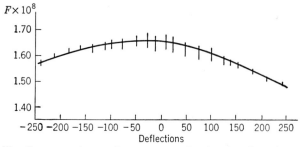

FIG. 5-4. Calibration curve for a galvanometer scale, showing the values of the figure of merit corresponding to different deflections. The lengths of the short lines show the uncertainties in the values of F.

5-6. Current Sensitivity of a Galvanometer. The figure of merit expresses the sensitivity of a galvanometer in the old way. The smaller the figure of merit, the more sensitive is the galvanometer. It is especially useful when the galvanometer is used to measure a small current, for then

$$i = Fd$$

The more modern practice is to express the sensitivity as the deflection D that is (or would be) produced on a scale at 1 m distance by a current of 1 μamp ($= 0.000001$ amp). This deflection is not observed directly but is computed, like F, from the observations in the above experiment. This gives

$$D = \frac{1}{10^6 F} = \frac{(P + Q)(R + g) + PQ}{10^6 PV} d$$

5-7. Megohm Sensitivity. The sensitiveness of a galvanometer is often expressed in megohms (millions of ohms). This means the number of megohms that can be placed in series with the galvanometer to give a deflection of one scale division per volt of the applied emf. Evidently this number is the same as D. In some respects this is a better way of expressing the sensitivity. A value of $D = 5,000$, for example, is the result of a computation and could not be actually observed, while it would be possible to place 5,000 megohms in series with the galvanometer and obtain a deflection of 1 mm when 1 volt is applied.

5-8. Voltage Sensitivity. For some purposes it is necessary to know the sensitiveness of a galvanometer to differences of potential applied to its terminals. This will depend largely upon the resistance of the galvanometer, as a small resistance will allow a larger current to flow.

The voltage sensitivity is expressed in volts per scale division and is given by the product of the figure of merit and the resistance of the galvanometer.

5-9. The Best Galvanometer. The best galvanometer to use for a particular purpose depends upon the special requirements. When the same current is passed through several galvanometers in series, the one that gives the largest deflection is the most sensitive for current measurements. Probably this one could have been picked out from the fact that it has the highest resistance because it is wound with many turns of fine wire. If the same galvanometers are joined in parallel and connected to a difference of potential of a few microvolts, probably a very different galvanometer will now show the largest deflection. This one will have a low resistance, and it will be sensitive for voltage measurements, provided that the source can supply the current required.

For many purposes a galvanometer may be "too good." If the current to be measured is not extremely minute, the deflection may be too large to measure on the scale. Extreme sensitiveness is often obtained by using a very fine suspension fiber. This makes the period long, and the zero, or resting point, is more variable than it is with a stiffer suspension.

For most purposes a galvanometer with D about 100 scale divisions per microampere and g about 100 ohms is satisfactory. The suspension should be strong enough to withstand moderate jars. The moving parts should have sufficient clearance to allow the use of the instrument without too tedious a process of leveling it. The interior should be easily accessible.

5-10. Factor of Merit. When the sensitiveness of a galvanometer is expressed in terms of the deflection D at a distance of 1,000 scale divisions, due to a current of 1 μamp, this sensitiveness can be increased by using a weaker suspension wire or by winding more turns of wire on the coil, without making any real improvement in the design of the galvanometer. In order to make a fair comparison of different galvanometers, it is necessary to take these differences into account.

To make this comparison it is usual to reduce the observed deflection D to what it would be if the period of the galvanometer were 10 sec and the resistance 1 ohm. It is found that D varies as T^2 and about as $g^{2/3}$, where T denotes the period of the galvanometer and g is its resistance.

The sensitiveness, thus modified, is called the "factor of merit" of the galvanometer. Thus

$$\text{Factor of merit} = \left(\frac{10}{T}\right)^2 \left(\frac{1}{g}\right)^{2/3} D$$

The unit of this quantity is not named. The unit of a similar quantity in which the reference period is taken as one second has been called a "D'Arson."

5-11. The Common Shunt. When the current to be measured by any instrument is larger than the range of the scale, the latter can be increased to almost any desired extent by placing a shunt in parallel with the instrument, as shown in Fig. 5-5. The shunt and instrument thus form two branches of a divided circuit, and the current through one branch is directly measured. When the current in this branch is known, the total current can be computed.

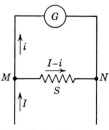

Fig. 5-5. Use of a shunt.

Thus let G denote the galvanometer or ammeter, and S the shunt. The fall of potential through the galvanometer from M to N is equal to the fall of potential through the shunt between the same points, or

$$ig = (I - i)s$$

where g denotes the resistance of the galvanometer and s that of the shunt. This gives for the value of the main current

$$I = i\,\frac{g + s}{s}$$

5-12. The Multiplying Factor of a Shunt. The factor $(g + s)/s$, by which the current measured by the galvanometer must be multiplied to give the total current in the main circuit, is called the "multiplying factor" of the shunt. In order that this factor may be expressed in convenient round numbers, 10, 100, 1,000, etc., it is necessary to have a series of shunts carefully adjusted to $\frac{1}{9}$, $\frac{1}{99}$, $\frac{1}{999}$, etc., of the resistance of the galvanometer. Such shunts will not have the same multiplying factor when used with a galvanometer of different resistance, and therefore they can be used advantageously only with the galvanometer for which they were made. Placing a shunt in parallel with a galvanometer reduces the total resistance of the circuit, and therefore the current measured by the galvanometer times the multiplying factor of the shunt does not give the value of the original current but the value of the new main current. Sometimes an extra resistance of $0.9g$, $0.99g$, or $0.999g$ is inserted to keep constant the total resistance of the circuit.

When the galvanometer is used ballistically, these shunt ratios are not the same as for steady currents, because of the varying amounts of damping produced by the different shunts.

5-13. The Universal Shunt. The *universal shunt* is so called because it can be used with any galvanometer and its shunt ratios will be the same. This arrangement is shown in Fig. 5-6. The total resistance, shown in the figure as 1,000 ohms, is connected as a permanent shunt on the galvanometer. The current to be measured divides at b, most of it passing through the part s of the total resistance S. In the figure s is shown as 10 ohms, the remaining 990 ohms acting merely as resistance in series with the galvanometer.

The expression for the multiplying factor of a universal shunt may be derived as follows: Let g denote the resistance of the galvanometer, S the total resistance of the shunt, and s the portion of S corresponding to a common shunt that carries that part of the current not flowing through the galvanometer. The fall of potential over the galvanometer and the part of S that is in series with the galvanometer is equal to the fall of

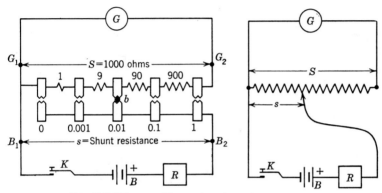

FIG. 5-6. Diagram illustrating the universal shunt.

potential over the part s. Let I denote the value of the main current and i the current through the galvanometer. Then

$$i[g + (S - s)] = (I - i)s$$

or
$$I = i\frac{g + S}{s}$$

For this form of shunt, therefore, the multiplying factor is inversely as s, since $(g + S)$ remains constant while s varies. For this reason, a shunt of this type can be used with different instruments, and the deflections of each will be proportional to the values of s for the same value of the main current. That is, the shunt will have the same series of multiplying factors for every galvanometer. The numbers $1, 0.1, 0.01$, etc. that are usually seen on a universal shunt box indicate the relative values of the galvanometer current i for the same current I in the main line.

5-14. Arrangement of the Universal Shunt. Any ordinary resistance box having a traveling plug for making a third connection at any intermediate point can be used as a universal shunt for any galvanometer, or two resistance boxes may be used when a shunt box as shown in Fig. 5-6 is not available. The shunt ratios are very accurate for all values, since all the coils are even ohms and can be adjusted much more precisely than with common shunts.   Differences in temperature between the galvanometer and shunt produce no error, but these differences should remain constant while measurements are being made.

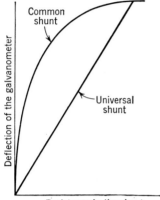

FIG. 5-7. Shows the relation between the galvanometer deflection and the resistance in the shunt.   With the same total current for each deflection, the straight-line curve for the universal shunt shows that the galvanometer deflection is proportional to the resistance in the shunt.

The curve for the common shunt shows the deflections of the same galvanometer with the same total current when common shunts of different resistances are used.   For small values of the shunt, the deflections increase rapidly as the shunt is made larger.   The larger deflections are not changed much by increasing the resistance of the shunt.

The advantage of the universal shunt is evident.

When a common shunt is used, the combined resistance of the shunt and the galvanometer is always less than that of the galvanometer.   With a universal shunt (especially when set at the 0.1 ratio) the combined resistance of the shunt and the galvanometer may be much larger than the resistance of the galvanometer alone.   In many kinds of work it is not essential that the resistance shall be constant or even known.   Where it must be known, it can be determined for the galvanometer and its shunt combined as readily as for the galvanometer by itself.

Universal shunts are marked with the numbers 0.1, 0.01, and 0.001, implying the fractions of the current that they pass through the galvanometer.   It is evident that when the universal shunt is used at the point marked 1, the galvanometer is in parallel with the large resistance of the entire shunt, and therefore the current in the galvanometer is not quite the full current on the main line.   If S is several times g, this slight reduction in the sensitiveness is of small moment.   The essential thing is that when the shunt is set at 0.1, 0.01, etc., the same total current will give deflections 0.1, 0.01, etc., large as with the shunt set at 1.   This arrangement of the galvanometer shunts is especially useful because a given series of shunts will have the same relative multiplying factors when used with any galvanometer.   Since the damping is constant, the shunt ratios remain the same when the galvanometer is used ballistically.

5-15. To Measure the Multiplying Factor of a Shunt. *Test of a Shunt Box.* The effect of a shunt in increasing the amount of current that can be measured by a galvanometer may be determined experimentally for the actual conditions of using the shunt.

Let us consider the case of a steady current I flowing through the circuit and dividing between the galvanometer and its shunt. The galvanometer will give a steady deflection d due to the current i through it. Then

$$i = Fd$$

where F is the figure of merit of the galvanometer alone.

Without changing the deflection or the current in any way, let us, secondly, think of the combined galvanometer and shunt as a single instrument. The total current I passes through this instrument, and the deflection is d. Then

$$I = F'd$$

where F' is the figure of merit of the combined galvanometer and shunt.

The multiplying factor m of a shunt is the ratio of the main current to the current through the galvanometer. This gives

$$m = \frac{I}{i} = \frac{F'd}{Fd} = \frac{F'}{F}$$

Thus, by determining the values of F and F', the multiplying factor of a shunt can be obtained. Or if m is known, the value of F' can be computed ($=mF$). This measured value of m should be compared with the computed value, $m = (g + s)/s$, when possible.

In the same manner, all the shunts in the shunt box should be tested and the multiplying factor of each one determined. The results should be compared with the values stamped on the shunt box and also with the computed values as determined by the relative resistances of the galvanometer and the shunt.

When using the shunts of lowest resistance, it may be better to determine the figure of merit by the direct-deflection method. The scale deflection should be about the same for each shunt.

5-16. Resistance of a Galvanometer by Half Deflection. (*a*) *Resistance in Series.* In some of the foregoing exercises it is necessary to know the resistance of the galvanometer as it has been used, either alone or combined with a shunt. When this is unknown, it can be determined with a fair degree of accuracy by the method of half deflection.

Let the galvanometer be connected to a source of potential difference that is small enough to keep the deflection on the scale with no additional resistance in series with the galvanometer. Then add enough resistance R in series with the galvanometer to make the deflection exactly one-half of its former value. This means that the current has been reduced to half its former value and therefore the resistance $R + g$ in the galvanometer circuit has now been made twice as much as when the galvanometer was used alone.   That is,

$$R + g = 2g \qquad \text{or} \qquad g = R$$

To avoid the errors arising from thermal currents, etc., it is best to reverse the battery and repeat the measurements, taking the mean of the two results as the correct value of g.   The resistance of a voltmeter can be measured by this method.

5-17. Resistance of a Galvanometer by Half Deflection.   (b) *Resistance in Parallel.*   With a sensitive galvanometer the arrangement shown

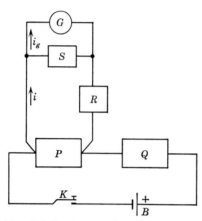

in Fig. 5-8 is used.   The potential divider PQ allows a small part of the emf of the battery to be used in the galvanometer circuit, and the large resistance R keeps the galvanometer current small enough to give a fair deflection, say 100 or 200 mm.   Then let a resistance box S be joined in parallel with the galvanometer and let the resistance in S be varied until the deflection is just half of its former value.   The current is now divided between the galvanometer and its shunt, half of the original current flowing through the galvanometer and the rest through the shunt.   If

FIG. 5-8. Resistance of a galvanometer. When $S = g$, $i_g = \frac{1}{2}i$.

the main current is unchanged, this means that the current is equally divided between the galvanometer and its shunt.   From this it follows that the resistance of the galvanometer is equal to the resistance of the shunt.

It is true that the addition of the shunt has reduced the resistance of this portion of the circuit, but as S is small compared with the resistance R in this circuit, the current i in the second case, which is divided between the galvanometer and the shunt, will be larger than the current in the first case by less than can be read on the galvanometer scale.   This point can be tested by placing the shunt resistance in series with R and noting whether it produces an appreciable effect in reducing the deflection.

5-18. A More Satisfactory Method.   *Full Deflection.*   It is not always true that half the deflection of a galvanometer corresponds to half as much current.   Naturally this casts some doubt on the accuracy of any half-deflection method.   This difficulty can be avoided by a slight change in the method shown in Fig. 5-8.   At the time the shunt S is added to the circuit, let the resistance P be doubled, keeping $(P + Q)$ the same as before.   This doubles the current through R, and this doubled current is halved by S, thus giving the same current in G as before. The deflection is at the same point on the scale as before, and no question arises regarding the relation of the two deflections.

5-19. Making an Ammeter. An interesting way to use a galvanometer is to construct a direct-reading ammeter with the galvanometer as the moving element. The problem is to set up the galvanometer with a suitable shunt and other resistances, so that when a current of 1 amp (or other assigned amount) is passed through the arrangement, the galvanometer will stand deflected 100 divisions.

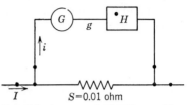

FIG. 5-9. An ammeter built up with a galvanometer and shunt.

When a known resistance of 0.01 ohm is available, it makes a useful shunt to carry most of the current that is to be measured. If the galvanometer is connected to the potential terminals, the current will divide between the two parallel circuits, most of it flowing through the low resistance of the shunt and a small part of it, i, flowing through the galvanometer.

From the law of shunts

$$I = \frac{g + s}{s} i = \frac{g + s}{s} F d'$$

where d' is the deflection corresponding to the main current I and F is the figure of merit of the galvanometer.

Since F and s are already fixed and we wish to have d' equal to 100 divisions when I is 1 amp, the resistance of the galvanometer circuit should be

$$g = G + H = \frac{I s}{F d'} - s = \left(\frac{1}{100F} - 1\right) s$$

If the galvanometer resistance is less than this, it can be increased by adding in series sufficient resistance H, as in Fig. 5-9, to give the desired amount.

If F has been determined accurately, the galvanometer should show a deflection of 100 divisions when the main current is 1 amp. This can be tested by putting a good ammeter in the main line to measure I. If this check is satisfactory, it gives confidence that by changing H and s the galvanometer can be set to measure any other desired range of current.

5-20. To Make a Microammeter. After an ammeter has been built and is found to measure the main current I as well as a good ammeter in the main line, an ammeter of any other range can be made with the same degree of accuracy.

Suppose it is desired to make an ammeter in which a main current of 1 μamp (0.000001 amp) will give a deflection of 100 divisions on the scale of the galvanometer. With this small current the value of the shunt S will need to be larger. Various values can be tried in the above formula for g to find resistances that are suitable and available to use for S and H. When the ammeter is built, one must trust the correctness of this computation for the accuracy of the instrument.

5-21. To Make a Voltmeter. A very excellent voltmeter can be made with a galvanometer and a high resistance of 100,000 ohms or more. The

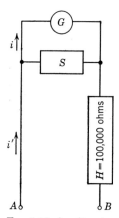

ideal voltmeter is an instrument that will measure emf without destroying or changing what it is trying to measure. Therefore, the less current that is drawn from the battery or other source, the better, since this will allow the available emf at the terminals to be nearer the value of the full emf.

Suppose the problem is to make a voltmeter that will read 100 divisions per volt. The arrangement of additional resistances depends upon the current sensitivity of the galvanometer, and this determines which of the following methods should be used.

By Series Resistance. The current through the galvanometer for a steady deflection of 100 divisions will be

FIG. 5-10. A voltmeter built up with a galvanometer and a high resistance.

$$i = 100F \tag{5-3}$$

where F is the figure of merit of the galvanometer. In order to obtain this current from a 1-volt source, the resistance of the galvanometer circuit must be

$$g' = H + g = \frac{V}{i} = \frac{1}{100F} \tag{5-4}$$

where g denotes the resistance in the galvanometer itself and H is the resistance that is added in series with the galvanometer.

Solving this equation gives the value of H. When this amount of resistance is available, it can be added to the galvanometer as shown in Fig. 5-10 (with S omitted) to make a direct-reading voltmeter. If H requires more resistance than is available, the following modification can be used.

By Shunt and Series Resistance. When the largest resistance that is available is still too small for H and would allow a current greater than

$100F$ in the galvanometer, a shunt S can be connected in parallel with the galvanometer, as in Fig. 5-10. By the law of shunts,

$$i' = \frac{g + s}{s} i \tag{5-5}$$

where i' denotes the value of the current in H and i is the part of this current that passes through the galvanometer. The total resistance of this circuit is

$$H' = H + \frac{gs}{g + s} \tag{5-6}$$

and the current for V volts between A and B is

$$i' = \frac{V}{H'} = \frac{V(g + s)}{Hg + Hs + gs} \tag{5-7}$$

Equating these two values for i' from (5-5) and (5-7) gives

$$\frac{V(g + s)}{Hg + Hs + gs} = \frac{(g + s)i}{s} \tag{5-8}$$

Solving (5-8) for the value of s gives

$$s = \frac{Hi}{V - (H + g)i} g \tag{5-9}$$

If it is desired to have a galvanometer current of $i = 100F$ with $V = 1$ volt, then the shunt should be

$$s = \frac{100FH}{1 - 100F(H + g)} g \tag{5-10}$$

Of course s cannot be computed any more closely than g is known. Hence the importance of knowing g is evident. Since the galvanometer coil is of copper wire, which changes with temperature, the value of g should be determined at the time it is used.

If this value of s is a fraction of an ohm over an integer and S can be varied only by 1-ohm steps, a little resistance can be added to g to make the value of s a whole number.

The voltmeter thus built can be compared with a standard voltmeter. When both voltmeters are in parallel with each other they should read the same.

To show the advantage of a high-resistance voltmeter, it can be used to measure the emfs of several old dry cells. After this is done, the same cells can be measured with the standard voltmeter. The latter will be found to give results that are too low, because of the internal resistance of the cells.

5-22. To Make a Voltmeter by Using a Microammeter.

If a microammeter has been constructed as described in Sec. 5-20, a voltmeter is readily made by adding a resistance of 1 megohm in series with it. Strictly, the total resistance of this voltmeter should be only 1 megohm, but if the microammeter is less than 1,000 ohms, the extra resistance will reduce the deflection by less than can be read on the scale.

This voltmeter can be compared with a standard voltmeter. It should read 100 divisions when in parallel with the standard voltmeter at 1 volt. This check tests both the accuracy of building the microammeter and that of making the voltmeter.

REFERENCE

Harris, Forest K.: "Electrical Measurements," John Wiley & Sons, Inc., New York, 1952.

PROBLEMS

5-1. The figure of merit of a certain galvanometer is 10^{-8} amp/mm. The scale is 1 m from the galvanometer. The coil resistance is 200 ohms.

a. Calculate the current sensitivity, megohm sensitivity, and voltage sensitivity.

b. What deflection would be produced by a current of 5×10^{-7} amp if the scale is moved to a distance of 75 cm from the galvanometer?

c. A 50-ohm resistance is placed in parallel with the coil. The scale is at 1 m from the galvanometer. Calculate the current sensitivity, megohm sensitivity, and voltage sensitivity for the shunted galvanometer.

d. The coil of this galvanometer is rewound. The area of the coil is not changed. The wire on the new coil has a diameter of one-half that of the original coil, but the total weight of the wire is not changed. Calculate the current sensitivity, megohm sensitivity, and voltage sensitivity.

5-2. *a.* The torsional constant (τ) of a galvanometer suspension is 50 dyne-cm/radian. Calculate the energy stored in the suspension when it has been twisted through an angle of 0.4 radian.

b. The moment of inertia J of the coil is 2 g-sq cm. Calculate the angular velocity which the coil will have when it passes through its equilibrium position if damping is neglected.

c. Neglecting damping, compute the period of oscillations of the coil.

5-3. The scale of a galvanometer is placed at a distance of 40 cm from the mirror. A deflection of 4.4 cm is observed. What is the angle through which the coil has turned?

5-4. Figure 5-4 shows a typical calibration curve for a galvanometer. Account for the variation of the figure of merit as the deflection increases.

5-5. An undamped galvanometer has a coil resistance of 500 ohms and a megohm sensitivity of 2,000. The external critical damping resistance of the coil is 1,200 ohms. Calculate the figure of merit, megohm sensitivity, and voltage sensitivity when properly damped.

CHAPTER 6

BALLISTIC GALVANOMETER

6-1. Introduction. The basic physical structures of the current galvanometer and ballistic galvanometer are quite similar. Their applications, however, differ widely. It has been seen that the current galvanometer is designed to produce a steady deflection which is proportional to the direct current passing through it. A ballistic galvanometer, on the other hand, is designed to give a deflection proportional to the quantity of charge which has passed through it. Since $q = \int i\,dt$, it is evident that the ballistic instrument is essentially an integrating meter.

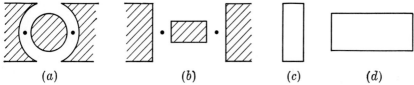

(a) (b) (c) (d)

Fig. 6-1. (a) Top view of a magnet used with a current galvanometer. A radial magnetic field is produced over a large region. (b) Top view of a magnet used with a ballistic galvanometer. Only the field in the region of the rest position is important. (c) Side view of a current galvanometer coil. (d) Side view of a ballistic galvanometer coil. The coil is made wide to give a large moment of inertia.

There are usually two features in a ballistic galvanometer which distinguish it from a current galvanometer. First, the moment of inertia (and therefore the period) is made large by using a wide coil. Second, the magnetic field need not have a radial shape. The relative shapes of a current and ballistic galvanometer are indicated in Fig. 6-1. The higher moment of inertia is important in that the charge to be measured must pass through the galvanometer coil before the coil has moved appreciably from its rest position. If this condition is not fulfilled, nonlinearities will result. In addition, since one measures the maximum swing when making measurements with the ballistic galvanometer, it is important to have a slowly moving coil so that the reading can be made accurately.

6-2. Calibration Constant of the Ballistic Galvanometer. The physical factors upon which the calibration constant of a ballistic galvanometer depends can be derived in the following manner: Let a transient current i

pass through the coil of the galvanometer when it is at its rest position; the passage of the current will produce a torque on the coil just as in the case of the current galvanometer.

$$\text{Torque} = KBNAi$$

The torque, acting during the short time that the transient current is flowing, will produce an angular impulse on the coil which will equal the change in angular momentum imparted to the coil.

$$\int KBNAi \, dt = J\omega$$

J is the moment of inertia of the coil. ω is the angular velocity with which the coil starts its swing after the passage of charge through it. Since $\int i \, dt = Q$, we have

$$Q = \frac{J\omega}{KBNA}$$

The coil thus moves with an angular velocity ω and an initial rotational kinetic energy

$$\tfrac{1}{2}J\omega^2 = \frac{Q^2(KBNA)^2}{2J}$$

As the coil rotates, it twists the suspension which produces a torque opposing the rotational motion. Eventually all the initial rotational kinetic energy of the coil will be used up in twisting the suspension, and the maximum deflection will be reached. At this point we can equate the initial kinetic energy of the coil to the work done in twisting the suspension.

$$\tfrac{1}{2}J\omega^2 = \int_0^{\theta_{\max}} \tau\theta \, d\theta$$

τ is the torsional constant of the suspension expressed in terms of the torque required to twist the suspension through a unit angle.

$$\frac{Q^2(KBNA)^2}{2J} = \frac{\tau\theta_{\max}^2}{2}$$

or

$$Q = \frac{(\tau J)^{\frac{1}{2}}}{KBNA}\theta_{\max}$$

As in the case of a current galvanometer, one commonly measures the deflection d of a light beam on a scale. Since d is proportional to θ, we can write $Q = kd$.

6-3. Damping of a Ballistic Galvanometer. The basic problems of damping the motion of a ballistic galvanometer are similar to those encountered with the current galvanometer. Several important differences must be considered, however. First, the ballistic galvanometer is not a steady-state instrument as is the current galvanometer. If the galvanometer is damped by a closed loop or by an external shunt (the $CDRX$), the damping effect will exist during the initial "kick" of the galvanometer and thereby reduce even the initial deflection θ_{max}. This usually will lead to a nonlinearity between the charge and the deflection produced. Second, the magnetic field of ballistic galvanometers is usually nonradial. This leads to a more complex equation of motion in which the damping force depends upon both the position of the coil and its velocity. Thus an ordinary damping system such as described in the previous chapter will not yield an ideal critically damped motion.

Two arrangements are commonly used to bring the coil to rest in a short time. A shunt may be placed across the terminals which, while not ideal, will produce a condition similar to critical damping. The value of this resistance may have to be determined experimentally. (This type of damping will, of course, reduce the sensitivity of the galvanometer.) A second and possibly more useful system is illustrated in Fig. 6-2. A key is connected across the terminals of the galvanometer and is normally open. This will have no effect on the initial kick of the galvanometer, thereby leading to a more linear relationship between deflection and charge. After the maximum deflection has been observed, the coil will swing toward its equilibrium position with its normal oscillating motion. If the key is depressed when the coil nears its equilibrium position, this overdamping effect will act as a strong brake, and by successively closing and opening the key, one can bring the system to rest at its zero position rather quickly.

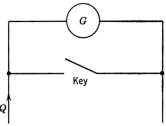

FIG. 6-2. Ballistic galvanometer with damping key. Closing the key produces a braking action on the coil.

6-4. Use of a Ballistic Galvanometer and Condenser. When the poles of a battery are joined to the plates of a condenser, the condenser becomes charged. The amount of this charge depends upon the emf E of the battery and the capacitance C of the condenser, being given by the relation

$$Q = CE$$

When the condenser is connected to a ballistic galvanometer, the charge Q in the condenser passes through the galvanometer and produces a deflection d, the relation being

$$Q = kd$$

where k is the constant of the galvanometer.

If this operation has been carefully arranged, so that there has been no leakage of the charge elsewhere, it is evident that the quantity that has been discharged through the galvanometer is the same quantity that was put into the condenser by the battery.   That is,

$$CE = kd$$

This is a very useful relation, since it can be used to determine the value of any one of the factors involved when the other three are known or can be measured.

The best arrangement for using a ballistic galvanometer with a condenser is shown in Fig. 6-3, where G represents the galvanometer and C the condenser, with the battery at B.   These are connected through the key as shown.   *Note that the tongue of the key is connected to the condenser and to the condenser only.*   This precaution is necessary in order that by no possibility can the battery ever be joined directly to the galvanometer. Arranged as shown, when the key is depressed, the condenser is joined to the battery and becomes charged.   When the key is raised, the condenser is joined to the galvanometer, through which the charge passes, producing a deflection.

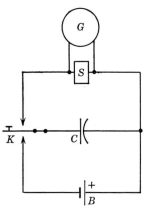

FIG. 6-3. Use of a condenser.

6-5. To Determine the Constant of a Ballistic Galvanometer.   For deflections that are not too large, the maximum swing is proportional to the quantity of electrons passed through the galvanometer, and the proportionality factor is called the "constant" of the galvanometer.   It is determined by discharging through the galvanometer a known quantity and noting the resulting deflection. Then

$$Q = kd$$

where k is the desired constant.   If k is known, for a galvanometer, any other quantity of electrons can be measured by sending it through the galvanometer and noting the corresponding deflection.

The galvanometer, condenser, and battery are connected as shown in Fig. 6-3.   The scale and telescope should be adjusted so that both the divisions and the numbers on the scale are distinctly seen in the telescope. The eyepiece must be focused on the cross hair of the telescope, which should appear very clearly defined.   When focusing the telescope upon the image of the scale as seen reflected from the mirror of the galvanometer, one must remember that the image is not on the surface of the mirror but lies as far back of the mirror as the scale is in front of it.

After the setup has been tried and found to work correctly with an old dry cell for B, a standard cell whose emf is known much more exactly may be substituted for it. Ten independent determinations of d should be made, with the resultant mean value used in the computation for k.

The relation given in Sec. 6-4 is

$$CE = kd$$

In the present case the condenser used must be one of known capacitance. Likewise, E must be known as exactly as the value of k is desired. And d is measured. Hence the computed value of k is

$$k = \frac{CE}{d}$$

If the value of C is given in microfarads, k will be expressed in *microcoulombs per scale division*. A microcoulomb is one-millionth of a coulomb and is the quantity of electrons that is represented by a current of one ampere flowing for one-millionth of a second.

The observations can be recorded as follows:

Galvanometer			Mean deflection at ____ meter	C	E	k
Zero	Reading	Deflection				

k for the scale 1 m from the mirror =

6-6. The Standard Value of k. When the mirror of a galvanometer is turned through a given angle, the magnitude of the corresponding deflection depends upon the distance of the scale from the mirror, being larger for the greater distance. This makes little difference for a given galvanometer with a fixed scale, but for the comparison of different galvanometers or for the same galvanometer with the scale at different distances, the value of the constant k has little meaning unless this factor is taken into account. Usually this is done by correcting the observed value of k to the value it would have when the deflection is read on a scale at 1 m from the mirror.

6-7. Comparison of EMFs by the Condenser Method. The arrangement of a condenser with a ballistic galvanometer may be conveniently used to measure the emf of a battery or any other difference of potential.

It thus serves as a voltmeter, with the advantage over the ordinary volt-meter that it measures the total emf of the battery, no matter what the internal resistance of the latter may be.

The setup is arranged as shown in Fig. 6-4. When the key is depressed, the condenser is charged, and when the key is raised, it is dis-charged through the galvanometer. It is immaterial whether the key works this way or the condenser is charged when the key is up and is discharged by depressing the key. It is abso-lutely necessary, however, that the tongue of the key be joined to the condenser, as is shown in Fig. 6-4.

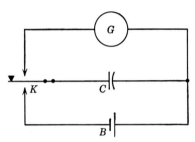

Fig. 6-4. To measure the emf of B.

When the battery is connected to the condenser, a current flows until the difference of potential between the two sets of condenser plates is equal to the emf of the battery. When this current is flowing, the available emf is only

$$E' = E - Ir$$

but as the charge in the condenser increases, the current through the battery becomes less and less, and when it reaches zero,

$$E' = E$$

Therefore the final difference of potential impressed on the condenser is equal to the total emf of the battery, and thus this emf can be measured.

The final charge in the condenser is

$$Q = CE = kd$$

Solving this for E gives

$$E = \frac{k}{C} d$$

The factor k/C can be determined by using a cell of known emf E_s and observing the corresponding deflection d_s. Then

$$E_s = \frac{k}{C} d_s \qquad \text{or} \qquad \frac{k}{C} = \frac{E_s}{d_s}$$

so that, finally,

$$E = \frac{E_s}{d_s} d$$

After this coefficient of d has been determined once and for all, the emf of any cell can be measured quickly and easily by observing the cor-responding deflection of the galvanometer.

Inasmuch as the reading must be caught quickly at the end of the swing, it will be best to take several trials and to use the mean deflection for computing the value of E. The readings may be recorded as follows:

Name of cell	Galvanometer			Mean deflection	$\dfrac{k}{C}$	Emf of cell
	Zero	Reading	Deflection			

6-8. Taking the Reading. The correct value of the deflection usually is not obtained on the first reading. The observer notes the end of the swing, perhaps to the nearest large division of the scale. Then on the second or third repetition of the deflection, knowing about where it will end, he takes the final reading to the nearest tenth of the smallest scale division.

It is better to read the position of rest (to the nearest tenth of a small scale division) than to try to set the scale on a given line, which usually results in other uncertainties. If the scale is moved at all, it should be by several divisions, so that the next observations of the zero point and the reading at the end of the swing will be taken on new parts of the scale and not be biased by an attempt to duplicate the first readings. The zero-point reading should be taken before the galvanometer is deflected. The difference between the readings at the end of the swing and at the zero point is called the "deflection."

6-9. Direction of Deflection. The galvanometer should be deflected a few times, always in the same direction, before it is finally read. With a sensitive galvanometer the coil does not return to exactly the same position of rest after it has been deflected in the opposite direction. After a few swings in the same direction, a condition is reached that can be duplicated for successive deflections.

6-10. Accuracy of Measurement. From an inspection of the equation $E = (k/C)d$, it is seen that the value of E that is computed by this formula depends directly upon the value of d. Therefore E cannot be determined any more exactly than d is known.

This means that the deflection should be determined as carefully as possible. The values of the capacitance of the condenser and the galvanometer shunt, if a shunt is used, should be so chosen that the deflection will be large enough to be measured accurately. For a deflection of 100 divisions, each division means 1 per cent of the whole. For a deflection of 10 divisions, each division is 10 per cent of the whole deflection, in which case the error due to the uncertainty in reading the deflection is very large.   Usually a deflection of 100 to 200 small scale divisions (millimeters) will be found suitable.   If it is too large, other uncertainties begin to appear.

When results better than approximate values are desired, several determinations of the deflection should be made, with the mean value used in computing the value of E.   If the individual deflections vary among themselves, this average value will be better than any single one.   But it should be noted that this average value has no meaning in decimal places beyond those estimated and recorded for each deflection.

6-11. Advantages of the Condenser Method.   One advantage of the condenser method is that it measures the full emf of the source, because no current is drawn and therefore there is no fall of potential in the resistance of the circuit.   It is true that while the condenser is being charged, there is a current through the battery, and writing the equation of potential differences for the charging circuit (Fig. 6-4) gives

$$E - ri - Ri - V = 0$$

where V is the voltage across the partly charged condenser and R includes all the resistances between the condenser and the battery of emf E and internal resistance r.

At the start, the charging current i may be fairly large, but the larger it is, the quicker will it become zero as the condenser becomes charged. At the end, when $i = 0$, the value of V is

$$V = E$$

and the final charge in the condenser is

$$Q = CV = CE$$

whatever the resistance of the emf source.

If the resistance in the charging circuit is very large, it may be necessary to keep the condenser connected to the emf source for some time—long enough to charge the condenser fully.   This time should be long enough that doubling the time will not increase the deflection of the galvanometer.

6-12. Effect of Resistance in the Charging Circuit.   It has been shown above that the condenser (Fig. 6-4) receives its full charge in a very short time, even with some resistance in the battery circuit.   Mathematically, it is true that the condenser never reaches its full charge; but as nearly as can be measured by reading the deflection of the ballistic galvanometer, the full charge is attained in a very short time.

How long is a "short time" for a given arrangement like Fig. 6-4? This question can be answered by observing the deflections when the key is closed on the battery side for different lengths of time, say for a fraction

of a second and for 1, 2, or 5 sec.   If doubling the time (with a good condenser) gives no larger deflection, then that time is long enough.

How much is "some resistance" in an arrangement like Fig. 6-4?   This
can be answered by adding various amounts, say 100, 1,000, 9,000, or
more, ohms in the circuit between the battery and the condenser.   If the
addition of a given resistance in series with the battery gives no observable decrease in the deflection, then this amount of resistance will cause
no trouble in measuring the emf of the battery.

Of course, these two questions are closely related.   When the resistance
in series with the battery is large enough to reduce the deflection by an
observable amount, then the key should be held closed for a longer time
in order to charge the condenser fully.   An old dry cell may have enough
internal resistance to require several seconds to charge the condenser.

6-13. Comparison of Capacitances by Direct Deflection.   The same
arrangement described above and shown in Fig. 6-4 can be used equally
well for the measurement of the capacitance of a condenser.   It is necessary only to go through the experiment as before and to observe the
galvanometer deflection with the first condenser, for which

$$CE = kd$$

Now, replacing the condenser by another one but using the same
battery and everything else the same as before, the relation becomes

$$C'E = kd'$$

where d' is the galvanometer deflection when the condenser of capacitance
C' is used.   Dividing the second equation by the first gives

$$C' = C \frac{d'}{d}$$

If C is a known capacitance, then the value of C' can be determined as
exactly as the flings d and d' can be measured.   Each of these should be
taken and recorded several times, with the mean values used in the
computation.

6-14. Calibration of the Scale of a Ballistic Galvanometer.   *Potential-
divider Method.*   The quantity of electrons discharged through a ballistic
galvanometer is measured by observing the deflection, or first throw, on
the scale.   The divisions on the scale are in millimeters or other equal
spaces, and the number of scale divisions passed over by the index gives
the magnitude of the deflection.   Sometimes the scale is bent into the arc
of a circle and the deflection measures 2θ, where θ is the angle turned
through by the mirror on the moving coil.   Most galvanometers are
designed to give deflections that, as nearly as possible, are proportional

to the quantity of electrons that is discharged through them. Just how nearly this is the case can be determined by calibrating the scale.

The calibration consists in discharging quantities of different known amounts through the galvanometer and observing the corresponding deflections. The different charges can be obtained by charging a condenser to different potential differences by the arrangement shown in Fig. 6-5. The battery B is connected to the potential divider PQ, and the current is maintained at a fixed value I by keeping the total resistance of $P + Q$ equal to a constant amount. By varying P while keeping $P + Q$ constant, various voltages PI can be obtained for charging the condenser.

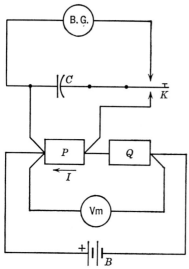

Fig. 6-5. Calibration of the scale of a ballistic galvanometer.

6-15. Plotting Values of k. From the relation derived in Sec. 6-4,

$$CE = CPI = kd \qquad (6\text{-}1)$$

from which

$$d = \frac{CI}{k} P \qquad (6\text{-}2)$$

This shows that the deflections as read on the scale should be proportional to the resistances in P when k, C and I are kept constant. However, it is very likely that k does not have the same value for deflections of various magnitudes. The arrangement of Fig. 6-5 enables one to examine this relationship. From Eq. (6-2), the value of k is

$$k = \frac{CPI}{d} = \frac{CPV}{d(P + Q)} \qquad (6\text{-}3)$$

where V is read on the voltmeter. The computation for k will be somewhat simplified if $P + Q$ is kept at 1,000 times V.

Using a 2-volt battery for B and varying P by 100-ohm steps, the observer can note the deflections and compute the corresponding values of k. If these 20 values of k agree with each other as closely as the accuracy of reading the deflections would lead one to expect, they can be averaged to give the mean value of k for the galvanometer.

But if the values of k show a continuous change as the deflections are made larger, it is better to draw a curve with the values of k for ordinates and deflections for abscissas. Since the variation in k is small, the curve can be magnified by taking the point $k = 0$ at quite a distance below the bottom edge of the sheet on which the curve is drawn.

From Eq. (6-3) it would appear that different deflections could be obtained by changing C as well as P. But in practice the values of C are not known as accurately as those of P. The condenser should be kept constant at such a value that the changes in P will give deflections over the full length of the scale that is to be calibrated.

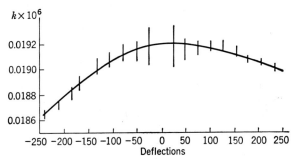

FIG. 6-6. Calibration curve for the scale of a ballistic galvanometer, showing the values of the constant k corresponding to different deflections. The lengths of the short lines show the uncertainties in the values of k.

In case the value of k as determined by the use of a standard cell does not lie on the curve of Fig. 6-6, the value that was used for V or for C in Eq. (6-3) should be modified so that the curve does pass through the point that was determined by the standard cell.

6-16. Internal Resistance of a Source by the Condenser Method. The condenser method offers a convenient and excellent means for determining the internal resistance of a cell, the principal advantages being the wide range of resistances that can be measured and the short time that the cell must be in use. In the ammeter-voltmeter method a considerable current must oftentimes be drawn from the cell and for

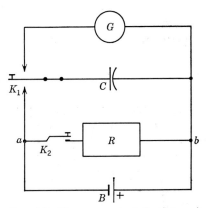

FIG. 6-7. Measurement of the internal resistance of a battery.

a period long enough to read both instruments. Such readings seldom can be repeated, for, owing to polarization, the cell does not return to its original condition.

The setup for using the condenser method is shown in Fig. 6-7. When K_2 is closed, there is a current through R and the cell, the value of which is

$$I = \frac{E}{R + r}$$

where r is the internal resistance of the battery. This gives

$$r = R\frac{E - E'}{E'} \tag{6-4}$$

where E' is written for RI, the external fall of potential. If R is known, only E and E' remain to be measured.

When K_1 is depressed, the condenser is charged to the emf E of the cell. This emf is measured by the deflection of the galvanometer when K_1 is raised, the relation being

$$E = \frac{k}{C}d$$

When a current is drawn from the cell by closing K_2, the potential difference between a and b is

$$E' = E - Ir = RI$$

If the condenser is now charged from this potential difference, the relation is

$$E' = \frac{k}{C}d'$$

Substituting these expressions in Eq. (6-4) gives

$$r = R\frac{d - d'}{d'} \tag{6-5}$$

When this method is used, it is necessary to keep K_2 closed only long enough to depress and raise K_1. With skill, this interval can be reduced to less than a second when the keys are worked by hand. If conditions require that the current be drawn from the cell for a shorter time than this, a special testing key can be used.

6-17. Special Testing Switch. An inexpensive switch that combines both K_1 and K_2 of Fig. 6-7 can be made from a double-pole, double-throw switch like the one shown in Fig. 6-8. The movable blade on one side is used for the tongue of K_1, and the two fixed jaws are the upper and lower contacts of K_1 (Fig. 6-7).

The movable blade on the other side of the switch and one of the fixed jaws serve as K_2. In the switch as shown in Fig. 6-8, both blades enter their jaws together. This means that K_1 and K_2 are both closed downward (Fig. 6-7) at the same time. Shortening the jaw of K_1 by a few

FIG. 6-8. The blades of this double-pole, double-throw switch can be used for K_1 and K_2 in Fig. 6-7 by shortening one clip. Care must be taken to have K_2 closed for only an instant. (*Courtesy of Leeds & Northrup Co.*)

millimeters will allow K_2 to make its contact first and to remain closed until after the condenser contact is broken. This assures that E' is measured, not E.

For the measurement of E the connection to R is opened by the key K (Fig. 6-9) and only the condenser side of the switch is used. In using this style of switch as a condenser key, great care must be taken not to touch any metal part of the switch with the fingers and thus partially discharge the condenser.

6-18. Diagram of the Switch. The connections that are made to the double switch of Fig. 6-8 are shown in Fig. 6-9, which is drawn similar to Fig. 6-7. H is the insulated handle that moves the blades of both K_1 and K_2. When in use, the blades are kept in the upper position, with the condenser connected to the galvanometer and its shunt S. Quickly throwing H down and back again

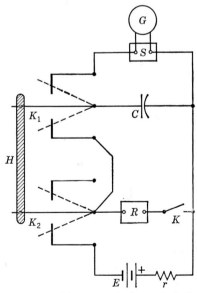

FIG. 6-9. Diagram of the switch in Fig. 6-8 and circuit in Fig. 6-7.

charges and discharges the condenser. If K remains closed throughout this operation, the voltage E' across R from a to b (Fig. 6-7) is measured. When K is open, the galvanometer deflection measures the full emf E of the battery E.

6-19. The Largest Uncertainty. It is seen in Eq. (6-5), above, that the value of r depends upon both d' and $d - d'$. When either one of these quantities is small, the percentage uncertainty in its measurement is large, and the value of r cannot be known to a better degree than the largest uncertainty in these factors. Therefore the resistance R should be set at a value that will make each of these quantities as large as possible.

Evidently both cannot be as large as d, since increasing one decreases the other. The uncertainty in $d - d'$ is due to the uncertainty in d as well as that in d'. Therefore, when these measurements are made, the value of R should be adjusted to give d' an intermediate value. This would be a value of R that will make d' between one-third and one-half as large as d. Thus neither d' nor $d - d'$ need be too small for accurate measurement.

Even with this optimum arrangement, the value of r is not determined any better than the largest uncertainty in d' or $d - d'$. The capacitance of the condenser should be sufficient to make these deflections as large as convenient.

6-20. Measurement of Insulation Resistance. A ballistic galvanometer can be used to advantage in the measurement of extremely high resistances. Usually such resistances need not be determined with great accuracy, but the order of magnitude is required. Resistances up to a few megohms, for example, the insulation of the electric wiring of a building from the water pipes, can be measured by a voltmeter. Larger resistances, such as the insulation of the coils from the frame of an electric motor, can be measured by the ammeter-voltmeter method with a sensitive galvanometer used in place of the ammeter.

But the current through the highest resistances, such as the insulation of electric cables or the resistance between the plates of a good condenser, is too small to be observed, even with a sensitive galvanometer. True, its value can be computed, but it is usually called a "leakage" current. It is then necessary to use one of the following methods.

6-21. Insulation Resistance by Leakage. This method is used when the resistance to be measured is so large that the current which it is possible to pass through it is too small to be measured by a sensitive galvanometer. The method consists, in brief, in letting the current flow into a condenser for a sufficient time and then discharging the accumulated quantity through a ballistic galvanometer.

The setup is arranged as shown in Fig. 6-10, where R denotes the large resistance to be measured. A battery of sufficient emf E supplies the current that flows through R and gradually charges the condenser C when K is closed on a. The switch K can be a charge-and-discharge key, or it can be one side of a double-throw switch. When a sufficient charge has accumulated in the condenser, it is discharged through the galvanometer by throwing K from a to b. The swings of the galvanometer can be checked by a shunt S that is adjusted to give critical damping.

At any instant during the time the current is flowing, the fall of potential over R is

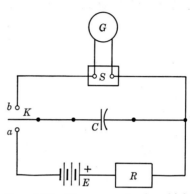

$$RI = E - V \qquad (6\text{-}6)$$

where V is the difference of potential across the condenser.

Fig. 6-10. Measurement of a high resistance.

In this method it is assumed that the condenser has considerable capacitance and that the condenser is discharged before V has reached an appreciable part of E. If the experiment is not worked in this way, the following discussion does not apply and the formula of Sec. 6-23 should be used.

At the start, and as long as V can be neglected in comparison with E, the current through R is

$$I = \frac{E}{R} \qquad (6\text{-}7)$$

If this current flows into the condenser for t sec, the accumulated charge is

$$q_x = It \qquad (6\text{-}8)$$

and when the condenser is discharged through the galvanometer, there is a deflection, or fling, of d scale divisions, such that

$$q_x = kd \qquad (6\text{-}9)$$

Thus the current is

$$I = \frac{kd}{t}$$

and

$$R = \frac{E}{I} = \frac{tE}{kd}$$

Determination of the Constant. The "constant" k of the galvanometer can be determined by the method described in Sec. 6-5, giving

$$k = \frac{E'C'}{d'}$$

where E', C', and d' denote the particular values used in this determination of the constant.

If the same battery is used in finding the constant as in the experiment proper, then $E = E'$, and the absolute value of the emf employed does not enter into the computation. The formula then becomes

$$R = \frac{t}{C'}\frac{d'}{d}$$

6-22. Comparison of High Resistances. If the value of one high resistance is known, other resistances of the same order of magnitude can be directly compared with it. This can be done without knowing the constant of the galvanometer or the capacitance of any condenser.

Suppose another resistance, R_1, is substituted for R in the setup shown by Fig. 6-10 and the current is allowed to leak through it into the condenser as before.

In the first case

$$R = \frac{tE}{kd}$$

and now

$$R_1 = \frac{t_1 E}{kd_1}$$

where t_1 and d_1 denote the observations made when the current was leaking through R_1.

Eliminating E and k by division gives

$$R_1 = R\frac{t_1 d}{td_1}$$

where, as in the preceding section, it is supposed that V can be neglected in comparison with E. This approximate value of R_1 becomes more nearly correct as d and d_1 are more nearly equal to each other.

6-23. The More Exact Formula. In Sec. 6-21 an assumption was made that is not quite true. It was there supposed that the leakage current remains constant, as shown by curve I (Fig. 6-11). From this assumption it follows that the charge collected in the condenser increases directly with the time the current is flowing or $q_x = It$, as shown by curve q_x (Fig. 6-12). As a matter of fact, the current becomes less and less as the condenser becomes charged, as shown by curve i (Fig. 6-11). The corresponding charge collected in the condenser is shown by curve q (Fig. 6-12). At the beginning there is not much difference between q_x and q, and it is seen that the method above is approximately true when t is a small part of the time required to half-charge the condenser.

In order to make a satisfactory measurement, it is necessary to collect a sufficient quantity in the condenser to give a measurable deflection when it is discharged through the galvanometer. Likewise, the time allowed for the collection of this quantity cannot be determined with accuracy if it is too short. Either or both of these conditions may make it desirable to allow the current to flow for a longer period and thus render it no longer permissible to neglect the increasing value of V.

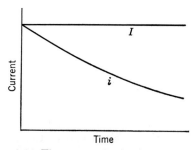

FIG. 6-11. The current flowing into a condenser decreases with time, as shown by the curve i.

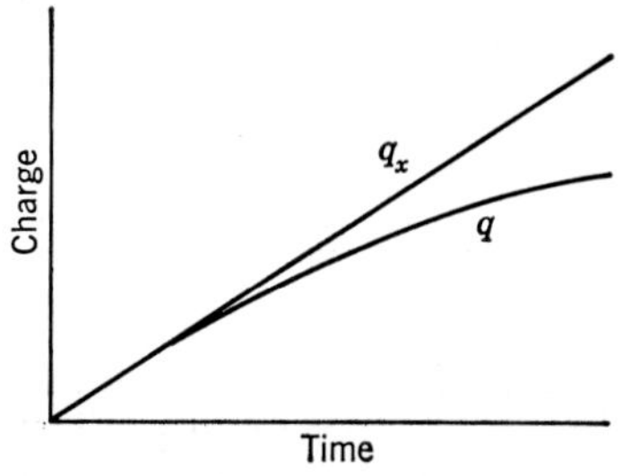

FIG. 6-12. The charge in a condenser increases with time, but more and more slowly, as shown by curve q.

Taking these conditions into account leads to the following considerations:

The current flowing into the condenser at any instant is

$$i = \frac{E - V}{R}$$

or, in terms of q, the charge in the condenser,

$$i = \frac{dq}{dt} = \frac{E - \frac{q}{C}}{R}$$

since $q = CV$.

Separating the variables for integration puts this in the form

$$\frac{-dq}{Q - q} = \frac{-dt}{RC}$$

where Q is written for EC. This Q is the charge in the condenser C when it is joined directly to E. Integrating this equation gives

$$\log (Q - q) = \frac{-t}{RC} + K$$

where K is the constant of integration.

At the start, when both t and q are zero, this equation gives

$$\log Q = K$$

After t' sec, when q' has collected in the condenser,

$$\log (Q - q') = \frac{-t'}{RC} + K$$

Subtracting the latter from the former gives the change during t' sec.

$$\log \frac{Q}{Q - q'} = \frac{t'}{RC}$$

Hence the value of R is

$$R = \frac{t'}{C \log \dfrac{Q}{Q - q'}} = \frac{t'}{2.303 \; C \log_{10} \dfrac{d_o}{d_o - d'}}$$

where d_o and d' are the deflections when the quantities Q and q' are discharged through the galvanometer.

6-24. Insulation Resistance of a Cable. Let a measured length of the cable be coiled up and placed in a tank of water to give contact over the outside surface if there is not a lead covering. Each end of the cable should extend out of the water or sheath for some distance and must be kept thoroughly dry. A few turns of bare copper wire are tightly wound around the cable near the middle of this dry portion of the insulation and are connected to the common point between the condenser and the battery, as shown in Fig. 6-13. This guard wire serves to intercept any current that may be leaking along the surface of the insulation. Connection is made to the water by means of a

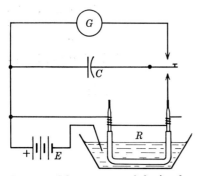

FIG. 6-13. Measurement of the insulation resistance of a cable.

wire of the same kind as that in the cable, to avoid voltaic effects. The wires connecting R to C and the key should be well insulated from the table and from all places where a charge can leak across. As far as possible, they should be stiff enough to stand in the air. Where necessary, they can be supported on blocks of sulfur.

The determination consists, then, in allowing the leakage current to flow into the condenser long enough (t' sec) to give a fair charge q', which, when discharged through the galvanometer, will give a deflection d'. When R is short-circuited and the condenser is charged directly from E, the charge is Q, giving a deflection d. For best results, q' should be about half of Q.

Owing to dielectric absorption of the charge, the first values of R will be too small. In reporting any value of R, the duration of the test should also be stated.

6-25. Insulation Resistance by Loss of Charge. If the high resistance also has considerable capacitance, it will not be necessary to use a separate condenser. Thus if it is required to measure the insulation of a condenser or of a long cable, the arrangement will be as shown in Fig. 6-14, where CR represents the condenser of capacitance C and resistance R. When K is closed, with K' also closed, the condenser is charged to the full potential difference of the battery. When the key is opened, the charge Q begins to leak away through the high resistance R. At the start, and before the charge in the condenser has been appreciably reduced by the leakage, the current through R is

$$I = \frac{E}{R}$$

This current will reduce the original charge in the condenser by the amount

$$q_x = It$$

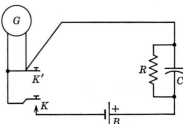

FIG. 6-14. Measurement of the resistance of a condenser.

in t sec, where t is not too great to consider the current constant during this interval. This loss of charge can be determined by recharging the condenser through the galvanometer to the original difference of potential. Then, as before,

$$q_x = kd \qquad \text{and} \qquad I = \frac{kd}{t}$$

from which

$$R = \frac{E}{I} = \frac{Et}{kd} = \frac{t}{C'}\frac{d'}{d}$$

where E cancels out if the same battery is used to determine k.

6-26. A More Exact Formula for the Loss-of-charge Method. As in the previous case, the current leaking out of C can be considered constant only as an approximation for a short portion of the discharge. For more precise results it will be necessary to take account of the decreasing value of this current.

As the charge in the condenser leaks away, the potential difference across the condenser decreases and finally becomes zero. Let V denote the value of this potential difference at the time t. The current flowing through R at this instant is $i = V/R$. Since this current is supplied from the charge in the condenser, it is measured by the rate of *decrease* of this charge, or

$$i = -\frac{dq}{dt} = \frac{V}{R} = \frac{q}{CR}$$

since $V = q/C$.

Separating the variables for integration,

$$\frac{dq}{q} = \frac{-dt}{CR}$$

and performing the integration gives

$$\log q = \frac{-t}{CR} + B \qquad (6\text{-}10)$$

where q is the charge remaining in the condenser after the charge has been leaking out for t sec.

At the start, when $t = 0$, $q = Q$. Putting these values in Eq. (6-10) gives, for the value of the constant of integration B,

$$\log Q = B$$

If q' denotes the quantity that has leaked out of the condenser in t' sec, then

$$q = Q - q'$$

and $$\log (Q - q') = \frac{-t}{CR} + \log Q$$

Solving this for R gives

$$R = \frac{t'}{C \log \dfrac{Q}{Q - q'}} = \frac{t'}{2.303 \, C \log_{10} \dfrac{d_o}{d_o - d'}}$$

where d_o and d' are the deflections when Q and q' are discharged through the galvanometer.

At the beginning of this test, the values of R will usually be too low

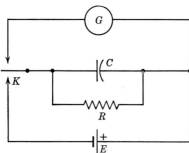

FIG. 6-15. The charge can leak away through the high resistance R.

because of the effect of "absorption," by which a part of the charge disappears. This reduces the charge in the condenser just as though it had leaked out. The true value of the resistance will be obtained only after several hours, in some cases several days; but if a first test is being made, it is well to determine the value of R at intervals of a few minutes. A curve plotted with the time of day for abscissas and the corresponding values of R for ordinates will show this variation and indicate the maximum value of the insulation resistance.

A resistance not having any capacitance can be measured by this method by adding a condenser in parallel with it. But in such a case the arrangement shown in Fig. 6-10 would be preferable.

6-27. To Reduce the Effect of Absorption. One difficulty in measuring the leakage resistance of a cable or a condenser arises from the absorption of the charge by the dielectric. This may be avoided in part by first charging the condenser and allowing it to leak for a period of t_1 sec. Next, discharge the remaining charge q_1 through a ballistic galvanometer and observe the deflection d_1. Then recharge and allow the charge to leak away for a longer period t_2, say twice as long. For the first period,

$$\log q_1 = \frac{-t_1}{CR} + B$$

and for the second,

$$\log q_2 = \frac{-t_2}{CR} + B$$

The difference is

$$\log \frac{q_1}{q_2} = \frac{t_2 - t_1}{CR}$$

and

$$R = \frac{t_2 - t_1}{C \log \frac{q_1}{q_2}} = \frac{t_2 - t_1}{2.303 \, C \log_{10} \frac{d_1}{d_2}} \qquad (6\text{-}11)$$

An electrometer can be used in place of the galvanometer, and it has the advantage that the loss of charge can be observed continuously. The curve of discharge can be plotted from readings taken at definite times. The effect of dielectric absorption of the charge is shown by the variation of this curve from the one plotted from Eq. (6-10). Any two points of this curve can be used in Eq. (6-11) to obtain the value of R.

6-28. The Quadrant Electrometer. The quadrant electrometer is an instrument for measuring electromotive forces or potential differences without drawing any current from the source that is being measured. In this respect it is more like a condenser than the ordinary voltmeter. It is better than the former in that it need not be discharged through a ballistic galvanometer; the voltage is measured by the steady deflection of the mirror as read on the scale.

The essential part of a quadrant electrometer is a small cylindrical box divided into four quadrants. Each quadrant is mounted on a pillar of amber or other good insulator, and the quadrants thus mounted are separated from each other by narrow air gaps (about 1 mm). Opposite quadrants are joined together by a light wire, and the two pairs of quadrants are connected to two binding posts, which also are well insulated from the case of the instrument. The object of this high insulation is to prevent the leaking away of any charges that may be given to the quadrants.

Within the hollow box formed by the quadrants is hung a large flat "needle" of thin aluminum or silvered paper. On the same axis is fastened a small mirror, and the whole is suspended by a flat strip of phosphor bronze or by a silvered fiber of quartz, in the same manner as a galvanometer coil. The upper end of the suspension wire is attached to a well-insulated binding post for outside connections.   The shape of this needle (horizontal plan) is such that as it turns there is little change in the size of the portion that is visible through the slits between the quadrants.

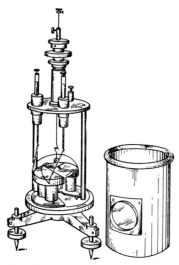

There is thus no change in the "edge correction" due to the nonuniformity of the electrostatic field near the edge of the quadrant.

When in use, the needle of the electrometer is connected through a high resistance to a battery of many volts. In the Dolezalek form shown in Fig. 6-16 this should be about 80 volts.   The other terminal of this battery should be connected to the metal case of the electrometer.   It is better to have this grounded if there is no other ground on the battery circuit.   The needle is thus given a charge, either positive or negative, but with the quadrants all at the same potential there is no deflection.   When a difference of potential is applied to the two

FIG. 6-16. A quadrant electrometer.

pairs of quadrants, the needle will turn into those quadrants having the greater difference from its own.   For a potential difference that is not too large, the deflection is directly proportional to this difference of potential; it is also proportional to the difference between the potential of the needle and the average potential of the quadrants.

A common form of electrometer is shown in Fig. 6-16.   At the center is seen the hollow brass box that is cut into four quadrants, with each quadrant mounted separately on its pillar of amber.   In this figure two of the quadrants have been drawn forward to reveal the suspended needle that hangs freely within this quartered box. When in use, these two quadrants are replaced in their symmetrical position.   A top view of the quadrants is seen in Fig. 6-17, and the outline of the needle is shown by the broken line.

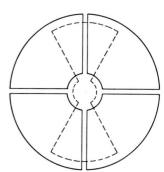

FIG. 6-17. Top view of the quadrants of an electrometer. The zero position of the needle is shown by the dotted outline.

6-29. Calibration of the Scale of a Quadrant Electrometer.   In order to determine the relation between the reading on the scale used with the quadrant electrometer and the corresponding potential difference between the quadrants, an arrangement such as that shown in Fig. 6-18 can be used. *A* and *B* denote the binding posts that are connected with the quadrants which are shown in section by Q' and Q. *D* represents the binding post that is connected to the moving needle through the suspension fiber.

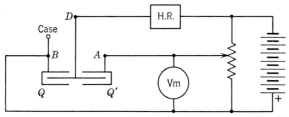

FIG. 6-18. Showing the arrangement for giving a known potential difference to the quadrants QQ'.

A potential divider of large resistance is connected across a battery of about 80 volts to give any desired part of this voltage between the quadrants. The value of this part of the voltage can be read from the voltmeter. The needle is connected through *D* to one terminal of the battery and thus is maintained at a constant potential above the case and one pair of quadrants. A high resistance is inserted in this line to avoid any accident if the needle should touch one of the quadrants.

The potential of the metal case of the electrometer is considered the zero from which the other potentials are counted. This arrangement of the connections is called "heterostatic," because the potentials of *A*, *B*, and *D* are all different. When the needle is connected to one pair of the quadrants instead of to the battery, the arrangement is called "idiostatic," because now the needle has the same potential as one pair of quadrants.

The calibration curve is obtained by setting the voltmeter at definite voltages about 5 volts apart and reading the corresponding steady deflections of the electrometer. The

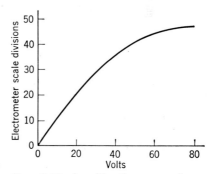

FIG. 6-19. A calibration curve for a quadrant electrometer connected heterostatically.

curve is plotted with the voltmeter readings as abscissas and the deflections as ordinates. For the smaller voltages this curve is nearly a straight line, but as the difference in potential between *D* and *A* becomes less, the deflection of the electrometer increases more slowly than at first.

6-30. Measurement of EMF with a Quadrant Electrometer. In order to measure an unknown emf with a quadrant electrometer, it is connected to the quadrants A and B (Fig. 6-18) with the needle still at 80 volts or whatever potential was used for the calibration curve. As soon as the needle settles down to a steady position, the deflection can be read, and the corresponding voltage is obtained from the calibration curve. No current is drawn from the source, and the measurement gives the full emf.

The difference of potential between the plates of a charged condenser can be measured in this way, since the condenser is not discharged by the electrometer.

6-31. Discharge Curve of a Condenser. When the plates of a charged condenser are connected to the terminals of a high resistance, the surplus electrons on the negative plate flow through the resistance until the lack of electrons on the positive plate has been filled. This electron current usually lasts for only a small fraction of a second, because the charges on the plates are always very small fractions of a coulomb. However, if the

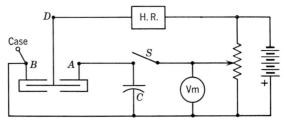

FIG. 6-20. Arrangement for continuous observation of the discharge of a condenser through its own high resistance insulation.

resistance of the circuit is large enough, the current will be very small and the condenser will be discharged slowly. With a resistance of several hundred megohms the condenser may discharge slowly enough to be followed with an electrometer. It is never possible to insulate the condenser plates so completely that there is no leakage of charge from them, and this leakage of a paraffined-paper condenser may be enough to discharge the condenser in an hour or so without any other conductor across the condenser.

In order that the condenser shall not be discharged by the handling of the connections, it should be connected to the electrometer beforehand, as shown in Fig. 6-20. The insulation of a key at S is not sufficient for this purpose. The connection at S should be a flexible wire hooked lightly to the terminal of the condenser. When this connection is opened, only that part which is still joined to the battery circuit should be handled, and a wide air gap should be made, leaving the condenser connected to the electrometer. By reading the electrometer every minute or two and plotting the readings against the time of day, a curve is obtained that shows the manner of discharge of the condenser.

A look at the calibration curve will probably show that there is not much change in the deflection for voltages from 80 down to about 50. Therefore in observing the discharge of the condenser it will be better to start with the condenser charged to about 50 volts.

When the charge in a condenser leaks out through a high resistance R, the current is

$$i = -\frac{dq}{dt} = \frac{V}{R} = \frac{q}{CR}$$

as shown in Sec. 6-26. The quantity q remaining in the condenser after the current has been flowing for t sec is

$$q = \epsilon^{-(t/CR)+B} = \epsilon^B \epsilon^{-t/CR} = Q\epsilon^{-t/CR}$$

from Eq. (6-10), Sec. 6-26, where Q is the maximum value of q.

The corresponding voltage across the condenser is

$$V = \frac{q}{C} = \frac{Q}{C}\epsilon^{-t/CR} = V_0\,\epsilon^{-t/CR}$$

and this is the equation of the curve shown in Fig. 6-21, when C and R are constants.

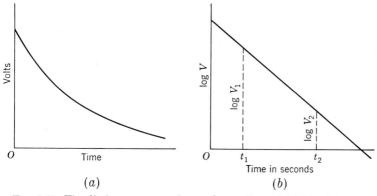

FIG. 6-21. The discharge curves of a condenser through high resistance.

6-32. Measurement of a High Resistance by an Electrometer and a Condenser.
If the charge on the condenser considered in the preceding article had leaked out through a larger resistance, the curve shown in Fig. 6-21 would have been less steep; and if the resistance had been smaller, the charge would have leaked out more rapidly. The form of the curve indicates, therefore, the resistance of the discharge circuit. Since it is difficult to draw a curve with accuracy, it is better to express this relationship as a straight line. This is done by plotting log V for the

ordinates of the curve instead of V. Since $V = q/C$,

$$\log V = \log q - \log C = B - \frac{t}{CR} - \log C$$

from Eq. (6-10). The constants B and $\log C$ can be eliminated from consideration by taking the difference between two values of $\log V$ that may be read from the straight line of Fig. 6-21b. The values of these are

$$\log V_1 = B - \frac{t_1}{CR} - \log C$$

and

$$\log V_2 = B - \frac{t_2}{CR} - \log C$$

giving a difference of

$$\log V_1 - \log V_2 = \frac{t_2 - t_1}{CR}$$

from which the value of R is given as

$$R = \frac{t_2 - t_1}{C(\log V_1 - \log V_2)}$$

If common logarithms are used in plotting the curve of Fig. 6-21b, the value of R is

$$R = \frac{t_2 - t_1}{2.303C(\log_{10} V_1 - \log_{10} V_2)} \qquad \text{ohms}$$

where C is the capacitance of the condenser in farads and $(t_2 - t_1)$ is the time in seconds during which the voltage across the condenser fell from V_1 to V_2 volts. If C is expressed in microfarads, the value of R will be given in megohms. The larger the resistance R, the less steep is the curve of Fig. 6-21.

Sometimes this curve of $\log V$ is not a straight line. This means that R has not remained constant during the time the curve was being taken. In this case the initial value of R can be obtained from the slope of the first part of the curve. The final value of R is given by using only the last part of the curve.

6-33. Accuracy of R. The expression for R is the reciprocal of the negative slope of the log curve of Fig. 6-21b. Therefore this curve should be plotted on scales that will make both $(t_2 - t_1)$ and $(\log V_1 - \log V_2)$ as large as possible on the plotted page. The value of R cannot be determined any better than these differences can be read from the curve. And the curve is no better than its agreement with the points through which it is drawn.

These considerations point out the place of the greatest uncertainty, and the more care should be taken to make this particular measurement with greater accuracy.

PROBLEMS

6-1. An amount of energy equal to $\frac{1}{2}(Q^2/C)$ is stored in a charged condenser. When the condenser is discharged through the coil of a ballistic galvanometer, a certain amount of mechanical energy $Q^2[(KBNA)^2/2J]$ is imparted to the rotating system. The efficiency of this arrangement for converting electrical energy into mechanical energy can be taken as the ratio of these quantities and is $[(KBNA)^2/J] C$. Is it possible to obtain an efficiency greater than unity by using a sufficiently large capacitor? Explain.

6-2. A 0.02-μf condenser is charged from a 5-volt battery. The condenser is discharged through a ballistic galvanometer immediately, and a deflection of 20 cm is observed. The process is repeated with the exception that the condenser is allowed to stand with a charge for 20 sec before discharging it. The deflection is now 18.5 cm. Compute the leakage resistance of the condenser.

6-3. A resistor is placed in parallel with the condenser of Prob. 6-2. After the condenser has again been charged from the 5-volt battery and allowed to stand for 20 sec, the deflection is found to be 17.5 cm. What is the value of the resistor?

6-4. An ideal condenser (0.01μf) is charged from a 2-volt battery in series with a large resistance. After charging for 100 sec the condenser is discharged through a ballistic galvanometer and a deflection of 6.3 cm is observed. The process is repeated, except that the condenser is allowed to charge for a long period of time. The deflection is now found to be 10 cm. Compute the value of the resistance.

6-5. An insight into the principle of operation of the electrometer or electrostatic voltmeter can be gained from a consideration of the following example: Parallel electrodes A and B, of length h and width y, are separated by a distance d. Plate A is held at a potential V, while plate B is at zero potential. A third plane electrode D, also of width y, is inserted to a distance x between A and B and is held at zero potential. Calculate the force which tends to pull D into the region. Neglect end effects. (*Hint:* Compute the energy W stored in the system as a function of x. Then $F = -dW/dx$.)

CHAPTER 7

MEASUREMENT OF RESISTANCE—
THE WHEATSTONE BRIDGE

7-1. Introduction. It is not uncommon to encounter resistances as low as 10^{-5} ohm or as high as 10^{15} ohms. It is evident that one cannot hope to cover such a large range adequately with a single instrument. The ammeter-voltmeter method has already been discussed (Chap. 3), and the limits of its usefulness can be obtained by considering the magnitudes involved. Let us suppose that voltages in the range of E volts can be measured accurately. We are now limited in the product of IR. In principle, low values of R can be measured if I can be made very large. This introduces a difficulty in that large amounts of heat are developed in the resistor with a resultant change in resistance. If one can safely dissipate P watts of power, the limiting lower value would be set at $R = E^2/P$.

The upper limit at which the ammeter-voltmeter method can be used is somewhat more flexible. If the current is measured by a milliammeter of reasonably low coil resistance, the limit V/R will be of the order of magnitude of the range of the current meter (10^{-3} amp for a 0- to 1-ma milliammeter). When the applied voltage is raised, the measurable resistance can also be raised by the same factor. Practical limits on the magnitude of this applied voltage are present, since at very high voltages surface leakage or a breakdown of the resistor due to sparking may take place. If we suppose that a potential difference of 100 volts is safe, a 0- to 1-ma milliammeter could be used in measuring resistances of the order of 10^5 ohms. If a more sensitive current meter, such as a galvanometer, were used, it might be possible to extend this limit to 10^9 or 10^{10} ohms without encountering breakdown and leakage problems.

7-2. The Portable Ohmmeter. The ammeter-voltmeter method is often used in the form of a convenient (although not very precise) portable ohmmeter (Fig. 7-1). The basic elements consist of a milliammeter, a source of voltage, and a series resistance R. The source of voltage is a battery with a potential-divider arrangement. In use, the test leads are shorted out and the potential divider is adjusted so that the meter needle corresponds to the highest reading on the dial. This reading is marked

0 ohms. This adjustment assures that the voltage in the circuit is the proper fixed value, and it eliminates the need for an actual voltmeter. The resistance to be measured R_x is now inserted between the test leads, and the meter deflects to a position determined by the current

$$i = \frac{V}{R + R_x}$$

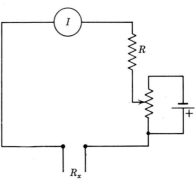

The scale can be marked off directly in ohms rather than milliamperes. It is clear that the scale will be non-linear; the position of half deflection of the meter will correspond to a value of the unknown resistance, $R_x = R$. Although the precision of such instruments is not extremely good, they should find wide use in the laboratory in determining the approximate value of a resistance before a more precise measurement by other methods is undertaken. The Wheat-

Fig. 7-1. Portable ohmmeter. The voltage is adjusted to give full-scale deflection with the terminals shorted out.

stone bridge furnishes one of the more precise methods for measuring resistances in this intermediate range of values (1 to 10^7 ohms).

7-3. The Wheatstone Bridge. The Wheatstone bridge consists, essentially, of two circuits in parallel and through which an electron current can flow. Let these circuits be represented by ABD and ACD (Fig. 7-2)

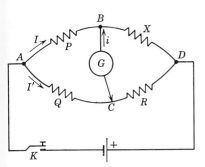

Fig. 7-2. Principle of the Wheatstone bridge.

and let the currents through the two branches be denoted by I and I'. Since the fall of potential from A to D is the same whichever path is considered, there must be a point C on one circuit that has the same potential as any chosen point B on the other, and a galvanometer bridged across between these points will indicate zero deflection. When the current in the galvanometer is zero with currents in the other branches, the bridge is said to be balanced.

To find the relations that will give the balanced condition, let us write out the equations of potential differences for the circuits of the network shown. Let the arrows indicate the direction that is *up* the potential gradient, *i.e.*, from the negative side of the battery toward the positive side. When the bridge is not quite balanced, let us take the case for which B is at a slightly higher potential than C.   Then as our pencil point traces over the circuit $ABCA$, the potential differences it passes over are

$$+PI - gi - QI' = 0$$

where g and i refer to the galvanometer.

Similarly, for the circuit $BDCB$,

$$+X(I + i) - R(I' - i) + gi = 0$$

where the current in X is the sum of the currents in P and the galvanometer.

For the special case of $i = 0$, when a balance is obtained, the equations become

$$PI - QI' = 0$$

and

$$XI - RI' = 0$$

Solving the first equation for I' and using this value in the second equation gives

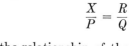

$$\frac{X}{P} = \frac{R}{Q}$$

as the relationship of the resistances when the bridge is balanced.   In the usual method of using the Wheatstone bridge, three of these resistances are known and the value of the fourth is easily computed from the above relation as soon as a balance is obtained.

7-4. Influence of the Battery Voltage and Galvanometer Sensitivity upon the Balance of the Wheatstone Bridge.   In the above

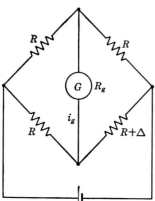

Fig. 7-3. Wheatstone bridge. A residual current i_g produces an uncertainty Δ in the determination of the resistance.

discussion we have assumed that the current through the galvanometer is truly zero when the bridge is balanced.   In practice, the experimental condition is somewhat different.   The resistances in the bridge are adjusted until no current flows *as indicated by the galvanometer*.   Since a galvanometer is not infinitely sensitive, an uncertainty will always be present in balancing the bridge.   We shall attempt to estimate the fractional error introduced by such uncertainties by considering the circuit of Fig. 7-3.   Assume that the bridge indicated has been balanced experimentally.   This means that the current through the galvanometer i_g is below the limit of detection for the instrument.   The uncertainty in the current through the galvanometer will introduce an uncertainty or error Δ in the value of the unknown resistance.   Applying Thévenin's theorem we find that the fractional error $\Delta/R \sim [i_g(R + R_g)]/E$. If we assume further that the minimum detectable current i_g is about as large as the figure of merit F for the galvanometer, this can be written $\Delta/R \sim [F(R + R_g)]/E$. For low values of resistance (that is, $R \ll R_g$), $\Delta/R \sim$ voltage sensitivity$/E$. For high values (that is, $R \gg R_g$), $\Delta/R \sim FR/E$. These relationships should be considered as "rule-of-thumb" criteria in choosing the proper galvanometer and battery for use in a given Wheatstone bridge measurement.

7-5. The Slide-wire Bridge—Simple Method. The Wheatstone bridge principle is used in several forms of apparatus for the measurement of resistance. The simplest of these is the slide-wire bridge as shown in Fig. 7-4. The unknown resistance that is to be measured is placed at X, while at R is the known resistance, usually a box of coils. The branch ACD consists of a single uniform wire, usually a meter[1] in length, stretched alongside or over a graduated scale. This wire should be of a

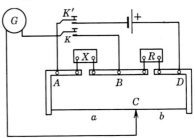

FIG. 7-4. The simple slide-wire bridge.

high-resistance alloy. The balance is obtained by moving the contact C along the wire until a point is found for which the deflection of the galvanometer is zero when K' and K are closed. This contact should not be scraped along the wire; it should always be raised, moved to the new point, and then gently but firmly pressed into contact with the wire. Neither should it be used for a key, as the continual tapping will dent the wire and destroy its uniformity. The two keys, K' and K, are combined into a single successive-contact key, often called a "Wheatstone bridge key," in which one motion of the hand will first close the battery key and then, after the currents have been established, will close the galvanometer circuit. The need of such a key is very evident when there is self inductance in X.

When the point C has been located, we have, as in Sec. 7-3,

$$xI' = apI$$

where a is the length of the bridge wire from A to C and p is the resistance per unit length of this wire. Thus ap is the resistance of this portion of the bridge wire, and x denotes the value of the resistance of X. Similarly,

$$RI' = bpI$$

where b is the length of the bridge wire from C to D. In this discussion the resistances of the heavy straps are neglected and A and D are con-

---

[1] Because a meter scale of 1,000 numbered divisions is available.

sidered as being at the ends of the bridge wire. Dividing each member of the first equation by the corresponding member of the second equation gives

$$\frac{x}{R} = \frac{a}{b}$$

Thus the ratio of the unknown resistance to R is given by the ratio of the two lengths into which the bridge wire is divided by the balance point.

This relation can be expressed in terms of a single length a by writing

$$b = c - a$$

where c denotes the total length of the bridge wire. Then

$$x = R\frac{a}{c - a} = R\frac{a}{1,000 - a}$$

if the total length of the bridge wire is 1,000 mm.

Measure in this way the resistances of two or more coils. Also measure the same coils when joined in series and compare the result with the computed value

$$R = R' + R''$$

When two coils are joined in parallel, the measured resistance should fulfill the relation

$$\frac{1}{R} = \frac{1}{R'} + \frac{1}{R''}$$

7-6. Calibration of the Slide-wire Bridge. In deducing the formula for the slide-wire bridge, it was assumed that the bridge wire was divided into 1,000 parts of equal resistance and that the readings obtained from the scale corresponded to these divisions. To make sure that the scale readings do thus correspond to the bridge wire, it is necessary to calibrate the wire, *i.e.*, to determine experimentally the readings on the scale that correspond to the 1,000 equal-resistance points on the wire.

Two well-adjusted resistance boxes are inserted in the two back openings of the bridge, and the battery and galvanometer are connected in the usual manner. Only the battery key is needed in this calibration, as good resistance boxes are noninductive. Closing the galvanometer key often gives a small deflection due to thermal emfs, and this deflection interferes with the determination of the balance point. It is better, therefore, to keep the galvanometer circuit closed and to observe the deflection when the battery circuit is opened and closed. If now, for example, 500 ohms is put in each box, the balance point will be at the middle of the bridge wire, and this should be at the point marked 500 on the scale. If it is not here but falls a distance f below 500, then f is the correction that must be added to the observed reading to obtain the true reading. Since there may be thermal currents in the galvanometer circuit, this value of f should be computed from the mean of two readings, one taken with the battery current direct and the other taken with the battery reversed.

In the same way the true location of the points 100, 200, 300, 400, 500, 600, 700, 800, and 900 can be found. It will be found convenient to keep the sum of P and Q always constant at 1,000 ohms.

Finally, a calibration curve is drawn, with the readings of the scale for abscissas and the corresponding corrections for ordinates. The cor-

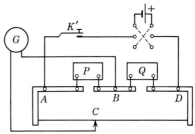

Fig. 7-5. Calibration of the simple bridge.

rections for any point on the scale can then be read directly from the curve. The corrected readings are thus expressed in thousandths of the total bridge wire, including all the resistance of straps, connections, etc., between the two points where the battery is attached.

CALIBRATION OF BRIDGE No. . . . .

P	Q	Scale readings			True reading $\dfrac{P}{P+Q}$	Cor.
		Battery direct	Battery reversed	Mean		
				(m)	(n)	$(n-m)$

7-7. Double Method of Using the Slide-wire Bridge.

In the simple method above, only a single balance point was obtained, and the value of the unknown resistance was computed from the relation

$$\frac{x}{R} = \frac{a'}{1,000 - a'}$$

where a' denotes the reading on the scale at the point of balance and is assumed to be the length of one portion of the bridge wire.

The measurement of resistance will be more precise if x and R are exchanged with each other, without, however, changing the value of either one, and a new balance point is determined. This second balance point will be, say, at a'' on the scale, and

$$\frac{x}{R} = \frac{1,000 - a''}{a''}$$

Combining these two equations by the addition of proportions,

$$\frac{x}{R} = \frac{1,000 + (a' - a'')}{1,000 - (a' - a'')} = \frac{1,000 + d}{1,000 - d}$$

in which the actual values of a' and a'' do not appear, but only their difference. Thus all questions regarding the starting point of the scale or the wire are eliminated, and if d is small, any error made in its determination will have only a small effect upon the value of x as computed from this equation.

7-8. Advantages of the Double Method. When it is desired to use the slide-wire bridge with some degree of precision, several precautions are necessary in order to avoid the principal errors. Prominent among these are the effects due to thermal emf in the galvanometer circuit, to balance which it is necessary to set the sliding contact on the wire at a point somewhat to one side of the true balance point to obtain zero deflection of the galvanometer. If the scale is displaced endwise with respect to the wire, or if the index from which the readings are taken is not exactly in line with the point at which contact is made on the wire, the effect is much the same.

Let a' denote the observed reading on the scale, and $a' + w$ the true balance point, where w denotes the displacement of the reading due to the causes noted above. The actual value of w is unknown, but it is constant in amount and sign, at least while one set of readings is being taken. For a balance of the bridge with x and R in the positions shown in Fig. 7-4, we have

$$\frac{x}{R} = \frac{a' + w}{c - (a' + w)}$$

where c is the total length of the bridge wire expressed in the same units as a' and w—usually in millimeters.

Interchanging x and R gives a new balance at a'', and

$$\frac{x}{R} = \frac{c - (a'' + w)}{a'' + w}$$

Combining these two expressions by the addition of proportions gives

$$\frac{x}{R} = \frac{c + (a' - a'')}{c - (a' - a'')} = \frac{c + d}{c - d}$$

It is seen that w has been eliminated by this double method and that the only measured quantity appearing in the final expression is d, the length of the wire between the two *observed* balance points. The value of

c should be determined by the method given in Sec. 8-2, as this may be greater than the meter of bridge wire because of the added resistance of the copper straps and the connections at each end of the wire. The "bridge wire" really includes all the resistance from A to D. However, if d is small, a slight uncertainty in the value of c will produce a negligible error in the computed value of x. This means that R should be taken as near to the value of x as is convenient.

7-9. The Best Position of Balance. (a) *Simple Bridge.* The formula deduced in Sec. 7-5 for the value of a resistance measured by the simple slide-wire bridge is

$$x = R \frac{a}{c - a} \tag{7-1}$$

Suppose that the value of a can be read to within $\pm h$ mm of its true value. This uncertainty will be the same at one part of the scale as at any other, but its effect in the formula depends upon the value of a. Using a value of a that is uncertain by $\pm h$ gives

$$R \frac{a \pm h}{c - (a \pm h)} = x \pm f \tag{7-2}$$

where f denotes the corresponding error introduced into x. It is required to find the value of a that will make the effect of $\pm h$ as small as possible.

7-10. The Total Error. Subtracting (7-1) from (7-2) leaves

$$\pm f = R \left(\frac{a \pm h}{c - (a \pm h)} - \frac{a}{c - a} \right) = R \frac{\pm hc}{(c - a \mp h)(c - a)} \tag{7-3}$$

While the uncertainty h cannot be neglected in computing the value of x, in finding the small quantity f it is near enough to call $c \mp h = c$. Then

$$f = R \frac{hc}{(c - a)^2} \tag{7-4}$$

This value of f, given by (7-4), is the uncertainty in the computed value of x by (7-1) due to the uncertainty h in determining the proper value for a.

In the notation of the calculus, the same result is obtained. When h is small, f corresponds to dx, and h to da. Differentiating (7-1) gives

$$\frac{f}{h} = \frac{dx}{da} = R \frac{c}{(c - a)^2}$$

the same as before.

7-11. The Relative Error. The relative error in the value of x as given by (7-1) is usually of greater importance than the actual error in ohms. It is evident that an error of 1 ohm in a total of 10 ohms is a very different thing from an error of 1 ohm in 1,000 ohms. The relative error is the ratio of the actual error to the total quantity measured. Thus from

(7-1) and (7-4) the relative error e is

$$e = \frac{f}{x} = h\,\frac{c}{(c-a)a}$$

From this it appears that even the relative error is not the same for the same error in reading but that it depends upon the value of a. If this expression is examined for a minimum value of e,

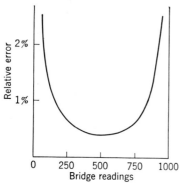

$$\frac{de}{da} = \frac{-c(c-2a)}{[(c-a)a]^2}\,h = 0$$

This is satisfied if

$$(c - 2a) = 0$$

Thus, in reading the value of a, a given error (say 1 mm) will produce the least effect on the computed value of x when the balance point comes at the middle of the bridge wire.

While this error is smallest at the middle of the bridge wire, it is not much greater when the balance falls at any point on the middle third of the scale. Figure 7-6 shows how rapidly this error increases when the balance point is near either end of the bridge wire.

FIG. 7-6. An uncertainty of 1 mm in estimating the value of a produces less error in x when the balance point is near the middle of the bridge.

7-12. The Best Position of Balance. (b) *Double Method.* The above discussion applies to the simple slide-wire bridge. When the double method is used, the formula for the resistance being measured is

$$x = R\,\frac{1{,}000 + (a' - a'')}{1{,}000 - (a' - a'')}$$

If this is written as

$$x = R\,\frac{c+z}{c-z}$$

it can be shown in the same way that a given error in measuring z will have the least effect upon the computed value of x when $z = 0$, i.e., when both a' and a'' are at the middle of the bridge wire. In this case

$$\frac{f}{h_z} = \frac{dx}{dz} = R\,\frac{2c}{(c-z)^2}$$

and

$$e = h_z\,\frac{2c}{(c-z)(c+z)} = \frac{2ch_z}{c^2 - z^2}$$

which evidently is a minimum when $z = 0$.

7-13. Sources of Error in Using the Slide-wire Bridge. These may be summarized as

1. Errors in setting, due to
   a. Thermal currents
   b. Contact maker not in line with index
   c. Nonuniform wire or scale
   d. Ends of wire and scale not coincident
2. Errors in reading, due to
   a. The position of balance
   b. True value of R, loose plugs, etc.

The effect of 1a and b can be eliminated by using the double method.
The only way to avoid the effect of c or d is to calibrate the bridge wire
and correct all readings, or to use the double method.

The error in reading the position of the index after a balance has been
found is often greater than the uncertainty of the setting.   In the preceding section it was shown that this error, which is about the same for all
parts of the scale, has the least effect on the computed value of x when the
reading is near the middle of the bridge wire.

The error in the resistance coils of a good box is very small.   However,
the value of R read from the box and used in computations may be very
different from the actual resistance of the experiment.   If some of the
plugs are loose or if they make poor contact because of dirt or corrosion,
the resistance may be considerably increased.   Moreover, the resistance
actually used in the bridge includes all the connections and lead wires
used to join the box to the bridge.   In the same way, the resistance measured includes the lead wires and connections.

If the apparent middle point of the bridge wire $[= \frac{1}{2}(a + a')]$ is not
constant for different measurements it may lead to greater uncertainty
than the errors mentioned above.

7-14. The Wheatstone Bridge Box.   In the slide-wire form of the
Wheatstone bridge, the balance is obtained by locating a certain point on
the wire, and the accuracy of the measurement depends upon the accuracy
with which the lengths of the two portions of the wire can be measured.
In the Wheatstone bridge box the wire is replaced by a few accurately
adjusted resistance coils.   Thus while the number of ratios that can be
employed is less than 10, the values of these few ratios are precise, even
when the ratio is far from unity.   The usual arrangement is to make P
and Q (Fig. 7-2) the two ratio arms with the unknown resistance in X and
to obtain the balance of the bridge by adjusting the resistance of R.   The
value of the unknown is then given by the usual relation

$$X = R \frac{P}{Q}$$

and is known as accurately as the value of R can be found.

In a common form of the Wheatstone bridge box, P and Q each contain

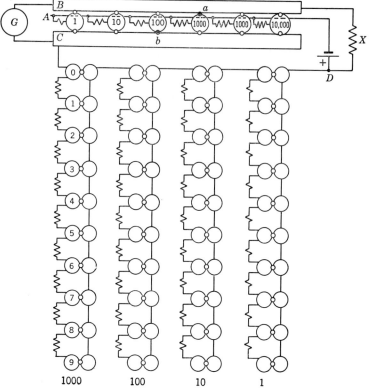

Fig. 7-7. Diagram of a decade Wheatstone bridge box.

1-, 10-, 100-, and 1,000-ohm coils, thus giving ratios of 1,000, 100, 10, 1, 0.1, 0.01, and 0.001. The rheostat arm R can be varied by 1-ohm steps from 0 to 11,110 ohms. This gives a range of measurement of unknown resistances from 0.001 to 11,110,000 ohms.

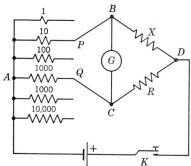

Fig. 7-8. Showing the six resistance coils from which P and Q can be chosen.

A more convenient form is the decade bridge. The rheostat arm is arranged on the decade plan with one plug for each decade. The resistance in this arm is indicated by the position of the plugs, which always remain in the box. The ratio arms consist of a single series of coils of 1 ohm and 10, 100, 1,000, 1,000, and 10,000 ohms, as shown in Fig. 7-8, and any coil can be used in either arm (but of course the same coil cannot be used in both arms at the same time).   The connections not visible are clearly indicated by lines drawn on the top of the box.   The different parts should be carefully compared with the diagram of Fig. 7-2, and the points A, B, C, D should be located before an attempt to use the box is made.   The resistance to be measured is joined to the posts marked X. The battery and the galvanometer are connected as shown, with a key in each circuit—preferably a successive-contact key.   When it is known that X is a noninductive resistance, it is better to omit the galvanometer key and thus avoid the troublesome deflections due to thermoelectric currents in the galvanometer circuit.

7-15. To Use a Wheatstone Bridge Box.   When starting to obtain a balance, set each of the ratio arms at 1,000 ohms and determine an approximate value of the resistance.   This is done by shunting the galvanometer with the smallest shunt available, and with R set at 1 ohm the keys are quickly tapped and the direction of the deflection is noted. The key should not be held down long enough to cause a large deflection, as the direction can be seen from a small one just as well and with less danger to the galvanometer.   Next, R is set at 9,000 ohms and the key is tapped.   Usually this deflection will be in the opposite direction.   If it is not, try zero and infinity.   Knowing that the value of X lies between 1 and 9,000, say, divide this range by next trying 100 ohms, and if X is less than this, try 10 ohms.   Suppose X is between 10 and 100.   This range is divided by trying 50, and so on until it is reduced to a single ohm.   Let us suppose that X is found to be between 68 and 69 ohms.   Then the ratio arms are changed so as to make R come to 6,800 or 6,900.   The exact value for a balance is determined by continuing the same process and is found, say, when R is 6,874 ohms.   This example then gives $X = 68.74$ ohms.

If the best balance, obtained with no shunt on the galvanometer or with a universal shunt set at 1.0, still gives some deflection, the next figure for X can be obtained by interpolation between two adjacent values of R, but this is not usually required.   If greater accuracy is desired, it is necessary to make a second measurement with the battery current reversed through the bridge.   This will reverse some of the errors, especially the effect of thermal currents in the galvanometer branch.   The mean of these two measurements will then be nearer the true value of X than either one alone.

Measure the resistance of several coils and check the results by measuring their resistance when joined in series and parallel.   If some of these coils have an iron core, notice the effect of first closing the galvanometer key and then closing the battery key.   Remember that the formula for this method was deduced on the assumption that all the currents were steady and that there was no current through the galvanometer.

The bridge balances can be recorded as follows:

Object measured	P	Q	R	X	Temperature of X

7-16. The Per Cent Bridge. A very useful and convenient form of the Wheatstone bridge is the arrangement shown in Fig. 7-9. When it is desired to compare coils that are nearly equal to each other, the ratio of the unknown resistance to the standard resistance is often more desired than the actual value of the unknown resistance. In the per cent bridge the coils are arranged so that the dial readings give directly the ratio of the unknown resistance to the resistance with which it is being compared.

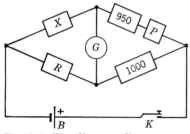

FIG. 7-9. The direct-reading per cent bridge.

As shown in Fig. 7-9, one of the ratio arms of the bridge is a fixed resistance of 1,000 ohms. The other ratio arm consists of a fixed resistance of 950 ohms in series with P, which is a dial resistance box of tens, units, and tenths of ohms. When X and R are equal, it will require 50 ohms in P to give a balance. Mark this setting of P as 100 per cent. If X is 1 per cent larger than R, it will require another 10 ohms in P, and this setting can be marked 101 per cent; 70 ohms would correspond to 102 per cent; etc. If the setting of P for a balance is 76.1 ohms, it indicates that X is 102.61 per cent of R. If the index on the first dial can be set to read "20" when it stands at 70 ohms, the last three figures, 261, can be read directly from the box. The range of this bridge is from 95 to 105 per cent. By increasing the ratio arms to 10,000 ohms each, the setting of P can be read to thousandths of 1 per cent. If lead wires are needed to connect X to the bridge, the lead wires for R should be made of the same resistance.

7-17. Resistance of Electrolytes. When there is a current through an electrolyte, it is accompanied by a separation of the substance in solution. The negative ions move in the same direction as the electron current, while the positive ions travel in the opposite direction, each being liberated at the electrodes. In general, this action causes polarization, which gives an extra emf in the circuit. In order, therefore, to measure the resistance of an electrolyte, it is necessary to employ an alternating cur--

rent. This can be most readily obtained from an audio-frequency oscillator.

The electrolyte is placed in a suitable cell and made the fourth arm of a Wheatstone bridge, the oscillator being used in place of the usual battery. The resistance of the electrolyte can then be determined by the Wheatstone bridge method in the usual way, and when the bridge is balanced,

$$r = R\frac{a}{b}$$

Since an alternating current is employed, this balance can be found by means of a telephone receiver connected in the usual place for a galvanometer. For purposes of instruction, the best form of cell for holding the electrolyte is a cylindrical tube with a circular electrode closing each end. The resistance measured by the bridge is then the resistance of the electrolyte between the two electrodes, and if the resistance of this column of the electrolyte is known, the resistivity s of the solution can be calculated the same as for metallic conductors, or

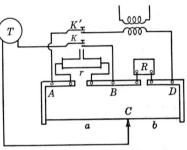

FIG. 7-10. Resistance of an electrolyte.

$$s = r\frac{A}{L}$$

where A is the cross section of the tube containing the solution and L is the distance between the electrodes. In case the form of the cell is such that A and L cannot be measured, the value of A/L can be determined by using a solution for which s is known.

The conductivity c of the solution is the reciprocal of this, or

$$c = \frac{1}{s} = \frac{L}{rA}$$

The resistance of an electrolyte, or, more strictly, its conductivity, depends upon the amount of the substance in solution—i.e., upon the number of ions per cubic centimeter. Therefore, if we wish to compare the conductivities of different electrolytes, it is necessary to express the concentrations in terms of the number of ions per cubic centimeter. This is usually stated in terms of the number of gram molecules of substance that are dissolved in 1 liter of the solution. For the purpose of this experiment it is necessary to express the concentrations in terms of the number of gram molecules in 1 cu cm of the solution. The molecular conductivity μ of an electrolyte is then defined as the conductivity per gram molecule of salt contained in each cubic centimeter of solution.

$$\mu = \frac{c}{m} = \frac{1}{ms} = \frac{L}{mrA} = \frac{bL}{amRA}$$

where m is the number of gram molecules in 1 cu cm of the solution.

The most interesting application of the conductivity of solutions is the knowledge it gives regarding the degree of dissociation of the dissolved substance. The conductivity of an electrolyte is due entirely to the ions it contains and is directly proportional to the number of ions per cubic centimeter. Most salts are completely dissociated in very dilute solutions, and therefore the molecular conductivity of such solutions is not increased by further dilution. Call this value μ_0. Then if μ denotes the molecular conductivity of a more concentrated solution of the same salt, the relative dissociation in the solution is

$$\alpha = \frac{\mu}{\mu_0}$$

Express results by means of a curve, using values of μ for ordinates and the corresponding values of $1/m$, which is the number of cubic centimeters containing 1 gram molecule, as abscissas.

PROBLEMS

7-1. A portable ohmmeter employs a 0- to 1-ma milliammeter as a detector. The mid-point on the scale should correspond to 5,000 ohms. What are the series voltage and series resistance for this instrument?

7-2. A Wheatstone bridge is set up using a 2-volt battery and a galvanometer which has a figure of merit of 10^{-5} amp/mm and a coil resistance of 1,000 ohms. What accuracy might be expected when measuring the following resistances: 10, 1,000, 10^4, 10^5 ohms?

7-3. Three resistances R_1, R_2, and R_3 are in parallel. Obtain the expression for the equivalent resistance R which can replace them.

7-4. If the values of the three resistances in Prob. 7-3 are known with an accuracy of ± 1 per cent, with what accuracy will R be known?

CHAPTER 8

THE WHEATSTONE BRIDGE

(Continued)

8-1. The Slide-wire Bridge with Extensions. The measurement of resistances by the slide-wire bridge can be made with more precision by using a longer bridge wire. The uncertainty in locating the balance points probably will be about the same, but since the distance $a' - a''$ between the two balance points is increased, the percentage error will be less.

As it would be inconvenient to have the apparatus much over a meter in length and as only the middle portion of the bridge wire is used in making careful measurements, the effective length of the bridge wire is increased by adding a resistance at each end. These extensions may consist of known lengths of wire similar to that used for the bridge wire—or any two equal resistances may be used and their equivalent lengths determined experimentally by the method shown below. The meter of wire provided with a scale then becomes only a short portion along the middle of the total length of the bridge wire. While this arrangement makes possible a greater precision of measurement, it also lessens the range of the bridge, as only those balances which fall on this limited section of the wire can be read.

The extensions are placed in the outside openings on the back of the bridge, between the ends of the bridge wire and the battery connections. They should be nearly equal. Let m' and m'' denote the number of millimeters of bridge wire having the same resistance as each extension, respectively, and let L denote the total length of the bridge wire including both of its extensions.

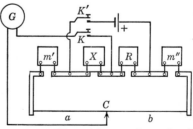

FIG. 8-1. Bridge with extensions m' and m''.

With the resistance X to be measured and the known resistance R in the middle openings of the bridge, as shown in Fig. 8-1, the first balance

121

point is found.   The reading on the scale at this point will be called a'.
Then

$$\frac{X}{R} = \frac{m' + a'}{m'' + b'} = \frac{m' + a'}{L - (m' + a')}$$

Interchanging the positions of X and R and calling the scale reading at
the new balance point a'',

$$\frac{X}{R} = \frac{L - (m' + a'')}{m' + a''}$$

And by the addition of proportions,

$$\frac{X}{R} = \frac{L + (a' - a'')}{L - (a' - a'')} = \frac{L + d}{L - d}$$

It is evident that this arrangement reduces the range of the bridge, for
only those values of X can be measured which are nearly enough equal to R
to give balance points on the scale.   But what is lost in range is more than
made up in the greater precision of measurement.

Dividing out the fraction gives

$$\frac{X}{R} = 1 + 2\frac{d}{L} + 2\frac{d^2}{L^2} + 2\frac{d^3}{L^3} +$$

All the terms after the second are negligible if d is small in comparison
with L, so that

$$X = R + 2R\frac{d}{L} \text{ (approximately)}$$

The only part of X, then, that is measured by the bridge is the second
term, and a small error in it will only slightly affect the computed value of
X.

8-2. To Find the Length of the Bridge Wire with Its Extensions.   The
total length of the bridge wire, including the extensions at each end, can
be determined as follows: A good resistance box P is used in place of the
unknown resistance X shown in the figure of the preceding section.   Then
with both extensions connected in the bridge, the values of P and R are
adjusted to bring the balance point near one end of the scale.   Let a'
denote this scale reading, corrected if necessary by the calibration curve
for this wire when used as a simple bridge.   Then

$$\frac{P}{P + R} = \frac{m' + a'}{m' + c + m''} = \frac{m' + a'}{L}$$

where c denotes the original length and L the total length of the bridge
wire.

Interchanging P and R, the balance falls near the other end of the wire, and

$$\frac{R}{P + R} = \frac{m' + a''}{L}$$

By subtraction,

$$\frac{P - R}{P + R} = \frac{a' - a''}{L}$$

whence

$$L = \frac{P + R}{P - R}(a' - a'')$$

This method may also be used to determine whether there is any extra resistance in the straps and connections at the ends of the usual meter of bridge wire.

Notice that the value of L can be determined only as accurately as the distance d ($= a' - a''$) can be measured. Therefore this distance should be as long as can be conveniently measured on the scale, say 85 or 90 cm.

8-3. To Calibrate the Slide-wire Bridge with Extensions. The formula deduced for this method works very well as long as X and R are nearly equal, but several errors may occur in its use, the principal of which are:

1. Using a wrong value for L
2. Neglecting all the terms containing d in powers higher than the first
3. Errors in the determination of d, due to nonuniform bridge wire, scale errors, etc.

The method of calibration described below corrects for all these errors at once by finding a correction to be added to the observed value of d, which will give to $1 + (2d/L)$ the true value of X/R.

With the bridge set up as shown in Fig. 8-1, with the extensions in place, and two good resistance boxes, P and Q, in place of X and R, we have

$$P = Q\left(1 + \frac{2d'}{L}\right) \tag{8-1}$$

and solving for d' gives

$$d' = \frac{L}{2Q}(P - Q) \tag{8-2}$$

Starting with P and Q each 1,000 ohms, the value of d should be zero. Then increasing P by successive small steps, the corresponding observed values of d can be determined. These observed values of d will not agree with the values of d' computed from Eq. (8-2) above, and therefore if used in Eq. (8-1), they will not give the correct values for P. This is because of the errors noted above. It is therefore necessary to add to the observed length of bridge wire d a certain amount h such that

$$d + h = d'$$

and this corrected value d' should be used.

In the present case where P and Q are known, the values of d' are computed from Eq. (8-2), while the corresponding values of d are observed on the bridge wire. The differences give the values of h, and a calibration curve can be drawn, as shown, that will give the correction to be used at each point. The correction increases rapidly with d, owing to the increasing importance of the second error noted above.

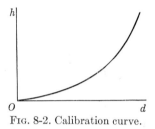

Fig. 8-2. Calibration curve.

In case d is not zero when P and Q are nominally equal, it means that one is really a little larger than the other. Let d_0 denote this value of d. This can be reduced to zero by adjusting the value of Q, or if it is more convenient, d_0 may be subtracted from each value of d throughout this calibration, as a sort of zero correction.

The readings may be recorded as below:

To Calibrate Bridge No. . . . With Extensions No. . . . and No. . . .

P	Q	a'	a''	d $= a' - a''$	d' Eq. (8-2)	h $= d' - d$

8-4. Measurement of Resistance by Carey Foster's Method.

One of the most exact methods for comparing two resistances is the one devised by Carey Foster of England. The Wheatstone bridge is arranged in the manner used for the slide-wire bridge with extensions, except that now the extensions become the resistances to be compared.

Thus in Fig. 8-3 let S be the resistance that is to be compared with R. These two are placed in the bridge as shown, being connected together

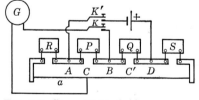

Fig. 8-3. Carey Foster bridge, comparing R and S.

by the bridge wire. The other arms of the bridge, P and Q, become merely ratio coils and may have any value, although they must be nearly equal to each other in order that the balance may fall on the bridge wire, and for the most sensitive arrangement, *all the arms* of a Wheatstone bridge should be nearly equal. For the greatest accuracy, R and S should each be in a constant-temperature oil bath.

Let the balance point be found by moving the contact C along the wire until a point is reached for which there is no deflection of the galvanometer. Let a' be the scale reading at this point. It is immaterial whether this scale extends the entire length of the bridge wire or not.

Now let R and S be interchanged. This will make no difference in the total length of the extended bridge wire ACD. But if the resistance of S is less than that of R, the new balance point will not fall at the same point as before, for it will be necessary to add to S enough of the bridge wire to make the part AC the same as before. The resistance of AC in the first instance is

$$R + a'p = \frac{P}{P + Q} T$$

After R and S have been exchanged, the resistance of AC' is

$$S + a''p = \frac{P}{P + Q} T$$

where T denotes the total resistance of ACD. Equating these two expressions,

$$R + a'p = S + a''p$$

and

$$S = R - (a'' - a')p$$

where p is the resistance per unit length of the bridge wire.

8-5. Correction for Lead Wires or Other Connections. If either S or R is connected to the bridge by wires or other connections, the resistance of these wires will be included in the measured values. If S and R denote the values of these resistances, exclusive of the lead wires, the last equation above should be written

$$S + s = R + r - (a'' - a')p$$

where s and r denote the resistances of the connections.

If S and R can be removed or short-circuited, leaving only the connections, a second pair of balances will give

$$s = r - (a_2 - a_1)p$$

Subtracting this from the first equation gives

$$S = R - [(a'' - a') - (a_2 - a_1)]p$$
$$= R - (d' - d_1)p$$

where d' and d_1 are the differences in the balance points in the two cases.

8-6. To Determine the Value of p. The resistance per unit length of bridge wire can be readily determined by the same arrangement as shown above. Let R and S be two resistances that are nominally equal. When the two balances are obtained,

$$S = R - (a'' - a')p$$

as shown above.

Now let S be shunted by a rather large resistance P so as to give a small but definite change in this resistance. Then

$$\frac{SP}{S + P} = R - (a_2 - a_1)p$$

where a_1 and a_2 are the new balance points. By subtraction,

$$S - \frac{SP}{S + P} = [(a_2 - a_1) - (a'' - a')]p$$

and

$$p = \frac{S^2}{(S + P)(d_1 - d')}$$

The resistances S and P should be well known, but extreme accuracy is not essential in order to obtain a fair degree of accuracy in the values of p.

8-7. The Ratio Box. An additional method for the exact comparison of two nearly equal resistances uses the ratio box shown in Fig. 8-4. The

Fig. 8-4. The ratio box. (*Courtesy of Leeds & Northrup Co.*)

principle of this method is shown in Fig. 7-9 (Sec. 7-16). The ratio box forms two arms of a Wheatstone bridge. When the bridge is balanced, the setting of the ratio box shows the magnitude of the unknown resistance in per cent of the standard. It can be used with resistors of the same nominal value, or in the ratio of 10:1. The range of the box shown in Fig. 8-4 is 99.000 to 101.110 per cent in steps of 0.001 per cent.

8-8. Comparison of Two Nearly Equal Resistances by Substitution. The Wheatstone bridge box, described in Sec. 7-14, is a fairly accurate instrument for the measurement of resistance. Its accuracy is limited by the accuracy of its own coils and the uncertainty in the connections to the unknown resistance. It is possible, however, to use such a box for measurements that are more accurate than its own coils.

Thus suppose that we wish to measure a resistance that is slightly less than 100 ohms and that there is available a standard 100-ohm coil whose resistance is accurately known. With the unknown resistance connected in the Wheatstone bridge, the latter is balanced as exactly as possible. If the best balance is not exact, the deflection of the galvanometer is carefully observed and recorded. Then without changing the bridge or moving any of the plugs, the standard resistance S is substituted in place of the unknown resistance X. If the key is now closed, the bridge will be found unbalanced, because, as we have supposed in this case, S is larger than X. When this is found to be the case, a high-resistance box R is connected in parallel with S and the resistance is adjusted until the galvanometer gives the same deflection as was observed for the best balance with X. Then the combined resistance of S and R must now be equal to X, or

$$X = \frac{SR}{S + R} = S\left(1 - \frac{S}{S + R}\right)$$

The actual values of the other arms of the bridge are thus immaterial.

An uncertainty in the resistance of the shunt R will affect only the value of the fraction $S/(S + R)$, which is small compared with 1. When X is nearly equal to S, thus requiring a large value of R to balance the bridge, the uncertainty due to R will be slight.

In case the unknown resistance is larger than the standard, the shunt should be applied to it. Then

$$S = \frac{XR}{X + R} \quad \text{and} \quad X = \frac{SR}{R - S} = S\left(1 + \frac{S}{R - S}\right)$$

8-9. Precise Comparison of Two Resistances. The last method leads directly to a method for the accurate comparison of two resistances that are nearly equal.

Let the two resistances to be compared form two arms S and R of a

Wheatstone bridge.   The other arms may be two resistances A and B of about the same nominal resistance as S.   The more nearly $A = B = S$, the better, but the exact values do not enter in the measurement.   The bridge is balanced by adjusting one or both of the high-resistance shunts P and Q that are in parallel with S and R, respectively.   When R is not closely equal to S, the balance may be obtained by the use of only one of the shunts.   If R is only slightly larger than S, the value of Q necessary to give a balance would be very high, possibly beyond the range of any available resistance box. In this case a moderate resistance can be used in P, making the resistance of this arm equal to

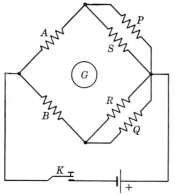

FIG. 8-5. The balance is obtained by adjusting the shunts P and Q.

$$S_1 = \frac{SP}{S + P}$$

It will then be possible to obtain a balance by adjusting Q, giving for the resistance of R and Q the value

$$R_1 = \frac{RQ}{R + Q}$$

When the bridge is thus balanced we have

$$\frac{A}{B} = \frac{S_1}{R_1}$$

where the ratio A/B is a constant but unknown quantity nearly equal to unity.

For the next step the resistances A and B are interchanged with each other without altering the rest of the bridge.   This, in general, will upset the balance of the bridge, but the latter can be restored by readjusting the shunts to new values P' and Q'.   We shall now have

$$\frac{A}{B} = \frac{R_2}{S_2}$$

where $\qquad R_2 = \dfrac{RQ'}{R + Q'} \qquad$ and $\qquad S_2 = \dfrac{SP'}{S + P'}$

Equating this to the former expression for A/B eliminates this unknown ratio, giving

$$\frac{S_1}{R_1} = \frac{R_2}{S_2}$$

or $\qquad \dfrac{R}{S} = \sqrt{\dfrac{PP'}{QQ'} \dfrac{(R + Q)(R + Q')}{(S + P)(S + P')}}$

In computing the value of this factor it is permissible to use the nominal values of S and R, since these are added to the large resistances P and Q.

The advantages of this method are apparent when it is seen that all uncertainty in the ratio of the arms A and B is completely eliminated, and since P and Q enter in both numerator and denominator, the uncertainties in their values will have only a slight effect on the result. Of course the usual precautions against thermal emfs and changes in temperature should be observed.

In some cases it is better to use the shunts on the ratio arms of the bridge, thus varying the ratio to equal R/S.

8-10. Measurement of Low Resistance. It is often necessary to measure a small resistance, in the form of either a short length of metal rod or a coil of low resistance. The determination of the resistivity of a metal usually requires the accurate measurement of a bar of the material.

The usual forms of the Wheatstone bridge are not well adapted for the measurement of such very small resistances. This is because of the uncertain resistances of the contacts and connections by which the small resistance is joined into the bridge. If the measured resistance is small, it is evident that a small contact resistance may introduce a relatively large uncertainty in the computed result.

8-11. Millivoltmeter and Ammeter Method. If a sensitive millivoltmeter is available, the resistance of a sample piece of wire MN (Fig. 8-6) can be measured by the ammeter-voltmeter method. As shown, the voltmeter measures the fall of potential between m and n. Since the main current does not pass through the contacts at m and n, the resistance of these contacts will not affect the fall of potential between m and n. These contact resistances are, in fact, a portion of

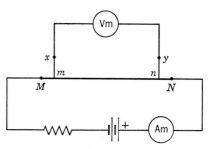

FIG. 8-6. Measurement of mn by the ammeter-voltmeter method.

the voltmeter circuit and therefore are negligible. The resistance measured by this method is, then, the resistance of that portion of the wire between m and n.

8-12. Four-terminal Resistances. If it is desired to preserve this particular sample mn as a fixed and constant resistance, the potential wires xm and yn should be soldered to the main wire MN at m and n. Then each time it is employed the same resistance mn will be used. The voltmeter is disconnected at x and y, and the current circuit is broken at M and N. All contact resistances in the measured portion of the wire are thus eliminated. Such fixed resistances are called "four-terminal resistances." Standards of low resistance are always thus provided with four terminals, two for current terminals and two for potential terminals.

8-13. Low-resistance Standard. An external view of a standard of low resistance is shown in Fig. 8-7, and the diagram of connections is given in Fig. 8-8. AB is a heavy piece of resistance metal of uniform

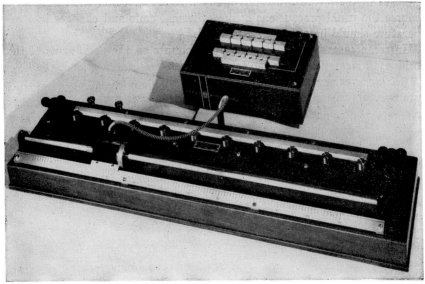

Fig. 8-7. Adjustable low-resistance standard. (*Courtesy of Leeds & Northrup Co.*)

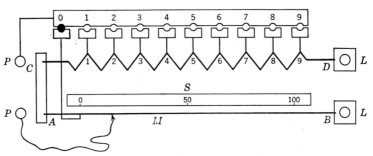

Fig. 8-8. Diagram of a variable standard of low resistance.

cross section and uniform resistance per unit of length; CD is another piece of resistance metal of smaller cross section, and the two are joined together by a heavy copper bar, AC, into which both are silver-soldered; LL are the current terminals and PP are the potential terminals. The resistance of AB between the marks 0 and 100 on the scale S is 0.001 ohm.

From the point 1 on the resistance CD to 0 on AB is also 0.001 ohm, from 2 to 0 is 0.002 and so on, and from 9 to 100 is the total resistance of 0.01 ohm. The slider M moves along the resistance AB and its position is read on the scale S, which is subdivided into 100 equal parts and can be read by a vernier to thousandths. Subdivided in this way the resistance between the tap-off points PP may have any value from 0.000001 to 0.01 ohm by steps of 0.000001 ohm.

There is a binding post on the bar AC that makes it possible to use AB separately. On account of its large current-carrying capacity, this bar is a very satisfactory standard resistance for measuring current by the potentiometer method.

8-14. The Kelvin Double Bridge. While low resistances can be meas-ured by the use of a millivoltmeter and an ammeter, the accuracy of the determination is limited by the accuracy with which the deflections of the instruments can be read. The best and most precise method for the measurement of low resist-ances is that of the Kelvin[1] double bridge, a modification of the Wheatstone bridge. As this is a null method, the results do not de-pend upon the readings of any instrument.

The general arrangement is shown in Fig. 8-9. The resistance to be

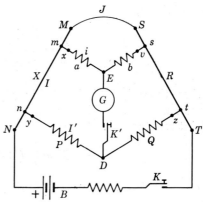

Fig. 8-9. Diagram of the Kelvin double bridge.

measured is shown by X, with current terminals at M and N, and poten-tial terminals at x and y. R is the standard low resistance with current terminals S and T and potential terminals v and z. R and X are joined at J with as good a connection as practicable, and together with P and Q they constitute the arms of a Wheatstone bridge. In the ordinary Wheatstone bridge the galvanometer is connected between D (Fig. 8-9) and the junction of X and R. In the present case this junction is not a single point but consists of the resistance $mMJSs$ with the unknown amount of contact resistance at M and S, and it is not possible to connect the galvanometer at any desired point along this irregular resistance. However, by using a potential divider ab in parallel from m to s, the galvanometer can be connected at a point E that has the desired potential between that of m and that of s. The potential at E is varied by adjust-ing the ratio of the resistances a and b.

Contact resistances at M and N do not enter as a part of X because they are not in the part being measured. At x and y the contact resist-ances are parts of the large resistances a and P.

The bridge can be balanced, giving zero deflection of the galvanometer,

[1] William Thomson, Lord Kelvin (1824–1907), British mathematician and physicist.

by changing the value of R or by varying the ratio of P to Q, whichever is the more convenient with the apparatus used.

Derivation of the Formula.   For this condition of balance, E and D will be at the same potential, and the fall of potential from n to D is equal to the fall from n to E.   Writing out this equation gives

$$PI' = XI + ai$$

Similarly, for the branches Dt and Et,

$$QI' = RI + bi$$

By division,

$$\frac{P}{Q} = \frac{X + a\dfrac{i}{I}}{R + b\dfrac{i}{I}} = \frac{X + ar}{R + br}$$

where the ratio i/I is denoted by r for convenience.   Evidently r will be a small number, and its smallness will depend upon the smallness of the joint resistance J.   Then

$$X = \frac{P}{Q}(R + br) - ar$$

$$= \frac{P}{Q}R + rb\left(\frac{P}{Q} - \frac{a}{b}\right)$$

If $P/Q = a/b$, this last term vanishes.   In case the equality is not precise, the effect of this term is made less by making r as small as possible. This means that the connection MJS, should have as low a resistance as possible.

8-15. Adjustment of the Ratios.   If the resistances that are used for a and b are not sufficiently accurate, the ratio a/b can be made actually equal to P/Q by slightly adjusting the value of a.   Opening the joint at J makes a Wheatstone bridge that will balance when

$$\frac{P}{Q} = \frac{X + a}{R + b}$$

X and R are usually small in comparison with a and b, and also are in the same ratio, so that the balance expressed by this equation can be made perfect by adjusting the value of a or b.   In this way a/b can be made to equal P/Q, and then a, b, and J do not appear in the final value for X.

When P/Q is set at 10^k, the result is given by the direct reading of R with the decimal point moved k places.

8-16. Advantages of the Kelvin Bridge.   Low resistances can also be measured by the potentiometer method.   With the Kelvin bridge, however, an absolutely steady current is not essential and the result is given by a simple computation from a single direct reading.

If for any reason it is necessary to use long wires to connect the bridge to the potential terminals of the low resistance, it will probably be better to use the potentiometer method.

8-17. The Student's Kelvin Bridge. The student's Kelvin bridge is a simplified instrument for making low-resistance measurements and for teaching the principles of these measurements. It consists of a 0.01-ohm calibrated variable standard of resistance, a double set of ratio coils with three different multiplying ratios, and a pair of current and potential contacts for the sample to be measured. It has, therefore, three ranges: 0 to 0.001 ohm, 0 to 0.01 ohm, and 0 to 0.1 ohm. It is designed for measuring samples about 18 in. long and up to $\frac{3}{8}$ in. in diameter, but binding posts are provided for connecting to any four-terminal resistance.

Fig. 8-10. Student type of Kelvin bridge. (*Courtesy of Leeds & Northrup Co.*)

As small a current should be used as will give the necessary sensitivity, and not more than 20 amp should be passed through the standard resistance. Excessive current will heat the wire, and its resistance will be increased. The zero reading of the galvanometer should be observed after the galvanometer has been connected to the bridge, to eliminate the effect of any thermal emf that may be present. The ratio of P to Q can be changed by moving one of the galvanometer connections near the center of the bridge. The other galvanometer connection controls the ratio of a to b. *Make both connections at the same value* (marked **10, 1,** or **0.1**), close the battery circuit, and balance the galvanometer by moving the contact on the standard slide-wire resistance. The maximum accuracy is obtained when the contact is at the upper end of the slide wire, and therefore a ratio value should be used that will bring the balance point as high as possible. The value of the unknown resistance X is obtained from the formula

$$X = R \frac{P}{Q}$$

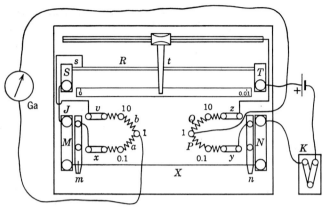

FIG. 8-11. Diagram of connections in the student's Kelvin bridge.

where R is the reading on the scale and the values of P/Q are stamped on the bridge.

The resistivity ρ of the metal is given by the relation

$$\rho = X \frac{A}{l}$$

8-18. Definite Relation between Resistance and Temperature. When the resistance of a wire is measured at different temperatures, it is found to have different values. Usually the resistance increases as the temperature rises. At each definite temperature, however, there is a definite resistance for each wire, and if corresponding values of resistance and temperature are plotted, the result will be a smooth curve that is not far from a straight line.

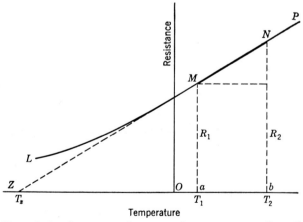

FIG. 8-12. The relation between resistance and temperature. Over the observed range MN, the resistance is proportional to the temperature above the point Z.

In alloys the change in resistance is very much less than it is for the pure metals, and in at least one case the resistance actually decreases with an increase of temperature.

Let $LMNP$ (Fig. 8-12) represent the relation between resistance and temperature for a given coil. Suppose the portion MN has been experimentally determined and plotted. If this limited part of the curve is nearly a straight line, the relation between the resistance R and the temperature T is easily expressed, as is shown below.

8-19. Supplemental Degrees for the Variation of Resistance with Temperature. The straight line MN shows a linear relationship between the resistance and the temperature of a given metal. This relation is not a direct proportionality, but it can be obtained by adding to each value of the temperature a certain constant that is characteristic of the particular metal considered.

Let R_1 and R_2 denote the resistances of a coil of metal wire at temperatures T_1 and T_2 on the centigrade scale. Then from the similar right-angled triangles ZaM and ZbN,

$$\frac{R_1}{R_2} = \frac{T_1 + T_z}{T_2 + T_z} \tag{8-3}$$

where the constant T_z is shown as ZO on the temperature axis. If this constant is to be added to T_1, it is evident that it must be expressed in the same units as T_1.

Solving Eq. (8-3) for the value of T_z gives

$$T_z = \frac{R_1 T_2 - R_2 T_1}{R_2 - R_1} \tag{8-4}$$

Thus it is not necessary actually to extend the line back to Z, since this point is readily determined by computation.

These formulas take no account of the change in dimensions with change of temperature and therefore apply to the conductor as a whole.

8-20. Determination of the Supplemental Degrees. In this experiment several coils of different metals are arranged in an oil bath where the temperature can be raised as desired. A very convenient arrangement is to use an electric water heater to hold the oil bath. With a suitable resistance in series this can be easily warmed and maintained at the desired temperatures. Starting at room temperature, carefully determine the resistance of each coil. A Wheatstone bridge box gives a convenient and accurate method for resistances having large temperature coefficients, like copper and iron. For alloys like German silver and manganin it is better to use a more delicate method, such as the Carey Foster bridge.

After the resistances of the coils have been obtained at room temperature, the bath is warmed 10 or 15°, and when things have become steady at the new temperature, the resistances are again measured. In the same way the resistances are determined at five or six different temperatures, and the results are plotted as a temperature-resistance curve for each coil.

The place on the temperature axis at which this portion of the curve points is determined by calculation, as shown above. In this computation it is best to use corresponding values of resistance and temperature as read from the curve and which therefore are less liable to error than are single observations.

TABLE 8-1. SUPPLEMENTAL DEGREES FOR THE RANGE 0 TO 100°C

Metal	T_z (centigrade)	Metal	T_z (centigrade)
Aluminum	236	Lead	243
Brass	556	Manganin, 0°–30°	40,000
Copper	234	Platinum	273
German silver	3,200	Silver	250
Gold	272	Tungsten	196
Iron (pure)	160	Zinc	249
Mercury	1,100	Platinoid	4,500

It is to be noted that the values of T_z in Table 8-1 imply nothing regarding the behavior of these metals at very low temperatures. This is merely a very simple way of expressing the slope of the temperature-resistance curve in the region where it has been studied.

8-21. Uncertainty. Probably the largest uncertainty in Eq. (8-4) is in the value of the small change $R_2 - R_1$ in the resistance. This difference, as read from the curve MN, contains all the error of the measured resistances. In plotting the observed values, a scale for R should be chosen that will bring M near the bottom of the graph sheet and N near the top.

The circumstances of the measurements will indicate the error in finding the temperature of the coil at the time its resistance is measured.

The observed values that are plotted for the curve MN will not lie exactly on a straight line. Therefore there will be some question as to the best location of the curve MN, especially as the plotted location of each point is more or less uncertain. Since the curve is a straight line, a ruler may be used to draw it. The observer must use his judgment as to how steep or how flat the line should be drawn. This variation in the position of the best line indicates the effect of all the uncertainties on the determination of $R_2 - R_1$.

8-22. Use of the Supplemental Degrees. When the resistance R_1 of a conductor at a temperature T_1 is known, it is a simple matter to find its resistance R at any other temperature T. Since ZMN is a straight line,

$$\frac{R_1}{T_1 + T_z} = \frac{R}{T + T_z}$$

The proportion shown in this equation is very convenient and can be used in many ways. It is especially useful when the change in resistance is large.

8-23. Resistance Thermometer. Oftentimes it is impossible to use a mercury thermometer in some situation where the temperature is desired. This may be an inaccessible part of some machine or a distant place in a factory. If a coil of copper or platinum wire can be used and its resistance measured, the temperature can be determined from the relation above.

One of the best methods for measuring temperatures that are either too high or too low for the use of mercury is to measure the resistance of a fine platinum wire and then read the temperature from the resistance-temperature curve. In fact, temperatures can be measured more exactly with a platinum resistance thermometer than with a mercury thermometer, even at ordinary temperatures.

8-24. Vacuum Measurements. The variation of resistance with temperature is also used in the measurement of moderately low pressures in vacuum systems. Two identical resistors in the form of fine wires are prepared in separate glass envelopes. One is evacuated to a very low pressure and permanently sealed off. The second is connected to the vacuum system. If the two wires are connected in series and a current is allowed to flow through them, a rise in temperature and resistance will result. The wire in the highly evacuated envelope can lose heat only by radiation and thus will come to equilibrium at a temperature T_1. The wire having its envelope attached to the vacuum system may lose heat both by radiation and also by "conduction" through the residual gas. Its equilibrium temperature T_2 will be lower than T_1 and will be a function of the gas pressure in the vacuum system. If these two wires are used as the series arms of a Wheatstone bridge, the pressure in the vacuum system can be obtained in terms of a resistance measurement.

8-25. Temperature Coefficient of Resistance. The change of resistance with temperature is often expressed in another way. From

$$\frac{R_1}{T_1 + T_z} = \frac{R}{T + T_z}$$

it follows that

$$R = R_1 \frac{T + T_z}{T_1 + T_z}$$

Clearing of fractions by performing this division gives

$$R = R_1 \left[1 + \frac{1}{T_1 + T_z} (T - T_1) \right]$$

$$= R_1(1 + a_1 t)$$

where t is written for the temperature difference $T - T_1$ and a_1 is written for $1/(T_1 + T_z)$. The coefficient a_1 is called the "temperature coefficient of resistance."

This formula is convenient to use in computing a small change in resistance corresponding to a small change t in temperature. Note that a_1 is the reciprocal of $(T_1 + T_z)$, where T_1 is the temperature at which the resistance of the conductor is R_1.

8-26. Second-degree Relation. In case a straight line cannot represent the resistance of a coil with sufficient accuracy, a curve showing this relation can be used. In this case all first-degree equations are likewise insufficient and it is then necessary to use a second-degree equation to approximate more closely the actual resistance-temperature curve. Usually this is written in the form

$$R = R_0(1 + at + bt^2)$$

where a and b are coefficients to be determined.

8-27. Temperature Coefficient of Copper. The effect of a small amount of other elements alloyed with copper appears not only in the decreased conductivity of the metal; the temperature coefficient is also decreased in the same ratio. Hard-drawing the metal has a similar effect. The temperature coefficient of a sample of copper wire expressed in terms of the resistance at 20°C is given by multiplying the number expressing the per cent conductivity by 0.00394.[1]

Table 8-2 gives a few values for copper furnished for electrical purposes and for the temperature range 10 to 100°C.

TABLE 8-2

Per cent conductivity	a_{20}	Supplemental degrees (centigrade)
101	0.00398	231.3
100	0.00394	233.8
99	0.00390	236.4
98	0.00386	239.1
97	0.00382	241.8
96	0.00378	244.6

PROBLEMS

8-1. A platinum resistance thermometer has a resistance of 10 ohms at 20°C. What is its resistance at 30°C?

8-2. Suppose that it is possible to measure a resistance change of 1 part in 10^5. What change in temperature could be measured by a platinum resistance thermometer?

8-3 What is the resistance of a coil of copper wire that increases 2 ohms per degree centigrade?

[1] *Bull. Natl. Bur. Standards*, vol. 7, p. 83, 1911.

CHAPTER 9

POTENTIOMETER AND STANDARD-CELL METHODS

9-1. A Scale for Voltage. When one wishes to measure an unknown quantity, like the height of a table or of a bush, it is necessary to have a standard scale to stand alongside the unknown in order to make the comparison. In somewhat the same way an unknown emf can be measured by comparing it with a scale of known emf. A simple way to make such a scale is shown in Fig. 9-1, where AD is a wire of high resistance and 2 m in length. In this wire is a current I sufficient to make the rise in potential from A to D exactly 2 volts. Then each millimeter along this wire will correspond to 1 mv, and a meter scale alongside the wire will measure the difference of potential from A up to any selected point.

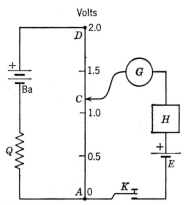

FIG. 9-1. A scale of voltage to measure the emf of E.

To use this scale let the negative side of the unknown emf, shown at E, be connected to the zero end of the scale at A. Let the positive side be connected through a suitable galvanometer to a movable key that can make contact with the wire at any point. When this contact is made too high on the scale, the galvanometer will deflect; when it is made too low, the galvanometer will deflect in the opposite direction. At the critical point where the rise of potential along the wire is equal to E, there will be no current in the galvanometer, and the value of E can be read on the millimeter (or millivolt) scale.

This method is better than one (as with a voltmeter) that draws current from E, because in this measurement there is no uncertainty due to the unknown fall of potential in the internal resistance of the cell that is being measured. The true value of E can be measured as accurately as the scale can be read.

9-2. Resistance-box Potentiometer. In the resistance-box potentiometer the slide wire is replaced by two similar and well-adjusted resistance

139

boxes P and R. Instead of actually moving the contact along the wire, resistance is transferred from one box to the other while the total resistance is kept constant (Fig. 9-2). In operation, a standard cell E_S is connected in the galvanometer circuit and P and R are adjusted so that no current flows through the galvanometer. Then

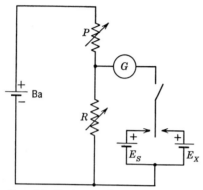

$$E_S = iR_S \qquad (9\text{-}1)$$

The unknown source of voltage E_X is next connected in the galvanometer circuit, and a new setting of P and R is made to again give zero current through the galvanometer so that

FIG. 9-2. Resistance-box potentiometer.

$$E_X = iR_X \qquad (9\text{-}2)$$

Since the sum of $P + R$ was held constant throughout the measurement, the current remains the same, and dividing Eq. (9-2) by (9-1) gives

$$\frac{E_X}{E_S} = \frac{R_X}{R_S} \qquad (9\text{-}3)$$

9-3. A Direct-reading Resistance-box Potentiometer. When many measurements of emf are to be made with a potentiometer, the computations indicated above may become very tedious. This work can be eliminated by adjusting the value of the constant current i to a simple round number. When

$$i = \frac{E_S}{R_S} = 0.00010000 \qquad \text{amp}$$

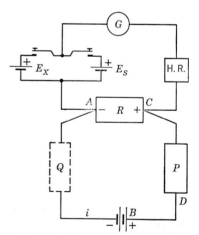

the computation consists merely in moving the decimal point four places to the left. This adjustment can be made as follows:

With a cell E_S of known emf (say 1.0183 volts), the resistance in box R (Fig. 9-3) is set to read this same number (say 10,183 ohms). This also fixes the resistance in box P, since $R + P$ is to be kept at some chosen and convenient value (say 19,999 ohms). This, in general, will not give

FIG. 9-3. Direct-reading potentiometer.

a balance, because Ri is too large.   Instead of changing R, reduce the current i by adding a third resistance Q (Fig. 9-3) in series with the main battery.   By adjusting the resistance in Q to give zero deflection of the galvanometer, the current i is brought to the desired value of 0.00010000 amp, where the four zeros at the right are known and significant figures.

With Q left at this value, the known cell E_s is replaced by the unknown cell E_X, whose emf is desired, and a new balance is obtained by adjusting R and P, keeping their sum constant at the previously chosen constant amount (say 19,999 ohms).

The emf of this cell is easily computed from the relation

$$E_X = R_X i = \frac{R_X}{10,000} \quad \text{volts}$$

where R_X is the resistance required for a balance.

Since the current in R and P is continually becoming less, its value should be determined near the time that it is to be used.

There should be no key in the circuit of PRQ and the main battery. If the battery B consists of one or two dry cells, the current i will be more nearly constant after it has been continuously flowing for a day or longer.

The readings can be arranged as below:

POTENTIOMETER MEASUREMENTS

Name of cell	R	P	Q	E

9-4. Standard Cells.

In all measurements of emf with the potentiometer it is necessary to have a known emf that can be used to standardize the value of the current through the potentiometer.   Such a known emf is furnished by a standard cell, which is a primary battery set up in accordance with definite specifications so that it will possess a definite emf. Such a cell is used as a standard of emf and is never expected to furnish a current.   Since the emf of a cell depends upon the materials used in its construction and not at all upon its size, standard cells are made small, both for economy of materials and for convenience in handling.

9-5. The Weston Normal Cell.

The Weston normal cell[1] has been the subject of much study and investigation, so that now it is possible for investigators in different parts of the world to set up such cells and know that they will have the same emf to within less than a ten-thousandth of

[1] Edward Weston (1850–1936), Anglo-American inventor.

a volt.   In order to attain such accuracy, it is necessary that the cells be set up in strict accordance with the specifications, using only the purest materials.

The cell is usually set up in an H-shaped glass vessel having dimensions of a few centimeters.   At the bottom of one leg pure mercury is placed to form the positive electrode of the cell.   Connection to this is made by a fine platinum wire sealed into the glass, the inner end being completely covered by the mercury.   At the bottom of the other leg there is placed, similarly, some cadmium amalgam, which, when warm, can be poured in like the mercury and then hardens as it cools.   A second platinum wire through the glass at the bottom of this leg makes electrical connection

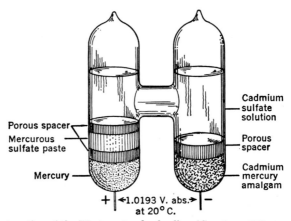

FIG. 9-4. Construction of the Weston standard cell.   (*Courtesy of Weston Instruments.*)

with the amalgam, which is the negative electrode of the cell.   The electrolyte is a saturated solution of cadmium sulfate, containing crystals of this salt in order to keep the solution saturated at all times.   The mercury electrode is protected from contamination by the cadmium in this solution by a thick layer of a paste consisting mainly of mercurous sulfate.   Cadmium ions from the solution coming through this paste form cadmium sulfate, and only mercury ions pass on and come into contact with the mercury electrode.   This paste is thus an efficient depolarizer.

The emf of a Weston normal cell that has been set up in accordance with the specifications is

$$E_t = 1.01830 - 0.0000406(t - 20°C) - 0.00000095(t - 20°C)^2$$
$$+ 0.00000001(t - 20°C)^3$$

This temperature formula was recommended by the London International Conference on Electrical Units in 1908, and the value 1.01830 international volts was found by the International Scientific Committee after an exhaustive series of measurements.

9-6. The Weston Unsaturated Standard Cell. The Weston normal cell described above is seldom used in ordinary electrical measurements. When it is kept at a constant temperature its emf is constant and definitely known, but it is not always convenient to maintain a constant-temperature chamber.

A considerable part of this temperature variation is due to the change in concentration as more or less of the crystals of cadmium sulfate dissolve. By omitting the crystals the electrolyte is unsaturated and remains at a constant concentration. Consequently, the emf changes very slightly with the temperature. The temperature coefficient is usually less than 0.000005 volt per degree centigrade and is almost invariably negative in sign. For most purposes the emf of the cell may be considered as constant for ordinary changes in room temperature. This modified form of cell is therefore more suitable for general laboratory use, although it is necessary to determine the emf of each cell that is made.

The construction of this Weston standard cell is illustrated in Fig. 9-4. The materials forming the electrodes are held in place by porcelain retainers provided with cotton packing. The cell is sealed to prevent leakage and evaporation.

To preserve the constancy of a standard cell, it should not be subjected to temperatures below 4 or above 40°C. If by chance it is subjected to extreme temperatures, it should be set aside for a month at a practically constant temperature. It will probably then be nearly back to its original emf.

9-7. Use of a Standard Cell. No current greater than 0.0001 amp should ever be permitted to pass through a standard cell, and then only for a moment. *A standard cell should never be connected to a voltmeter.* The internal resistance is high, and therefore it would not be possible for it to furnish much current. In fact, any appreciable current drawn from the cell would polarize it somewhat, thereby decreasing its emf by an unknown amount and thus destroying the only value of the cell. The depolarizer tends to restore the emf to its original value, but the time required depends upon the amount of polarization.

Standard cells may be used to charge condensers to a known difference of potential, for in this case there is no steady current drawn from the cell, and the transient current is not sufficient to cause an appreciable polarization. When used in connection with a potentiometer, the cell should always be placed in series with a sensitive galvanometer. If it is necessary to reduce the deflection, this should be done by means of a high resistance in series with the cell instead of by a shunt on the galvanometer, which would still allow the large current to flow through the standard cell. When using the galvanometer, the key should be tapped lightly and quickly so as to give a deflection of only a centimeter or two. This will indicate the direction of the current as clearly as a larger deflection and does not injure the standard cell.

9-8. The Student's Potentiometer.   For purposes of instruction the three-box potentiometer described previously enables one to see all parts of the setup, and as has been shown, it is convenient to use.   But if one has many measurements of emf to make, the mere changing of the resistances to obtain the balances becomes a laborious task.   It is then desirable to have a quicker and simpler way of varying the resistances.

In the student's potentiometer shown in Fig. 9-5 and the diagram (Fig. 9-6), the resistances corresponding to the slide wire of Fig. 9-1 are arranged in two dials.   Moving the sliding contacts over these dials corresponds

Fig. 9-5. Student type of potentiometer.   (*Courtesy of Leeds & Northrup Co.*)

to moving the ends A and C of the galvanometer circuit along the slide wire (Fig. 9-1).

There are no moving or sliding contacts within the potentiometer, except the two contact points at A and B (see diagram, Fig. 9-6).   All other connections are brazed or soldered.   The moving contact points at A and B are in the galvanometer circuit and not in a calibrated resistance circuit, so that any variation in contact resistance at these points can cause no error in the measurements.

Connections from the calibrated resistors are brought out to ten binding posts, five on each end of the instrument.   The five binding posts on the left of the instrument and the one in the central position on the right are the six used for potentiometer measurements.   The other four binding posts at the right of the instrument are provided for making independent connections to the slide wire, either with or without end coils, so that the instrument may be used as part of a Wheatstone bridge circuit.   For this purpose an additional scale 0 to 1,000 is provided on the slide wire.

To standardize the potentiometer against the standard cell, switch S is closed to STD CELL, contacts A and B are set to correspond to the certified cell emf, and the rheostat is adjusted until galvanometer G shows a balance as a key is closed.   The regulating rheostat is a four-dial resistance box, total 999.9 ohms, adjustable in steps of 0.1 ohm.   With a 2- or 3-volt battery, the total circuit resistance will be 200 to 300 ohms and the four-dial rheostat regulates the current within ± 0.05 per cent or better. As the current is only 0.0101 amp, dry cells are a satisfactory source.

Two tapping keys are shown: one for closing the circuit through a 10,000-ohm resistor P, to aid in making the initial balance and to protect

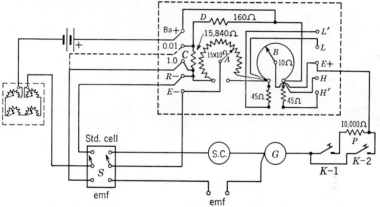

FIG. 9-6. Diagram of the electrical circuits in the student's potentiometer.

galvanometer and standard cell; the other to close the circuit directly to obtain the final balance.

The range is changed by two shunt coils C and D, so proportioned that when connection is made to binding post 1, the current through A and B is 100 times that through C and D.   When connection is made to post 0.01, coil C is in series with A and B and the current through this circuit is now $\frac{1}{100}$ of its former value.   The total current remains the same.

9-9. The Type K-2 Potentiometer.   A more elaborate form of potentiometer is shown in Fig. 9-7, and the electrical connections are shown in Fig. 9-8.   The essential part of the instrument consists of 15 five-ohm coils AD.   These are adjusted to equality to a high degree of accuracy and have in series with them an extended wire DB, the resistance of which from 0 to 1,000 on its scale (the entire scale reading from 0 to 1,110) is also 5 ohms.   A contact brush M can make contact on the studs in 5-ohm steps, and a contact point M' can make contact at any point on the extended wire DB.   Current from the battery Ba flows

FIG. 9-7a. Type K-2 potentiometer. (*Courtesy of Leeds & Northrup Co.*)

FIG. 9-7b. Type K-3 potentiometer. (*Courtesy of Leeds & Northrup Co.*)

146

through these resistances, and by means of the regulating rheostat Q it is adjusted to be 0.02 amp.   The fall of potential across any one of the coils AD is consequently 0.1 volt, and that across the extended wire DB is 0.11 volt.   By placing the contact point M' at zero and moving the contact M, the fall of potential between M and M' can be varied by steps of 0.1 volt from 0 up to 1.5 volts.   If the contact point M' is moved along the wire, the fall of potential between M and M' can be varied in infinitesimal steps.

In making measurements, the known emf—in practice the standard cell—is connected across one portion of AO and the current is adjusted to

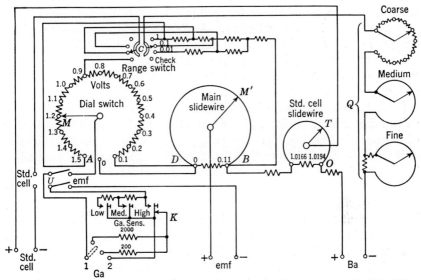

FIG. 9-8. Diagram of the electrical circuits in the potentiometer shown in Fig. 9-7a.

0.02 amp; the unknown potential difference can be balanced against the fall of potential in another portion of AO.   The connections for accomplishing this with a single galvanometer are shown in Fig. 9-8.   At the point 0.9 on the series of resistance AD, a wire is permanently attached that leads to one point of the double-throw switch U.   The resistance between 0.9 and B corresponds to the emf of 1.01 volts, and between B and 1.0166 is a sufficient addition to make the resistance between 0.9 and this point correspond to an emf of 1.0166 volts.   At this point there is a small circular slide wire with a movable contact point T through which connection to the standard cell is made.   The slide wire has a resistance value that makes possible a practically continuous variation in the standard cell circuit of voltages from 1.0166 to 1.0194.   This range corresponds to the variations among different cadmium cells.

This part of the potentiometer circuit is connected to the standard cell and galvanometer when the switch U is closed on the side marked STD CELL. When thus connected to the standard cell with T set at the value of the emf of the cell, the current through the potentiometer is adjusted by the rheostats Q until the galvanometer shows zero deflection. This makes the potentiometer direct reading in volts, and with the switch U on EMF, the brushes M and M' can be moved to the positions for a balance with the unknown emf.

When thus balanced on the unknown emf, the switch U can be tipped quickly to STD CELL to see that the adjustment of Q is still correct, then quickly back to EMF to see that the setting OM' is still right. If the galvanometer shows a balance for each position of U, the reading of MM' will give the emf of the unknown. A slight deflection calls for a slight readjustment of Q and a corresponding readjustment of M'.

Keys are provided in the galvanometer circuit with various series resistances so that excessive currents can be avoided during the initial adjustments of the potentiometer.

9-10. Galvanometers for Use with a Potentiometer. The function of a galvanometer in connection with a potentiometer is that of an indicator of absence of current and, consequently, of absence of potential difference at its terminals. It is essential, therefore, that it should respond readily by a deflection when a slight potential difference exists.

When the total resistance of the circuit containing the galvanometer is relatively low, a small unbalanced potential difference will produce a larger current than when the total resistance is high. In low-resistance circuits, as in ordinary voltage and current measurements, and in thermo-couple work, the instrument should have a good microvolt sensitivity, while in high-resistance circuits, such as those met with in many of the potential measurements of physical chemistry, an instrument of high current sensitivity is desirable. At the same time, the internal resistance and external critical damping resistance, in the first instance, should be low; in the second, these values should be correspondingly higher.

9-11. Calibration of a Voltmeter by the Potentiometer Method. A low-reading voltmeter can be readily calibrated by the potentiometer method described above. The voltmeter is connected to the sliding part of a potential divider so that the pointer of the voltmeter can be brought to any desired place on the scale. The resistance of the potential divider should be a few hundred ohms, or enough to draw only a small current from the battery so that the voltage will remain fairly constant. The voltmeter, while still connected with its battery, is also inserted in the galvanometer circuit of the potentiometer in place of the unknown emf. There will thus be two electric circuits, each with its own battery: one battery, W, supplying the current through the potential divider and the voltmeter, while the potentiometer current through P and Q comes from the battery B. The galvanometer circuit connects these two circuits, and the current through it may be in either direction or made zero by adjusting the resistance in R.

Writing the equation for the potential differences in the circuit through the voltmeter and the galvanometer, before a balance has been obtained, but with H. R. reduced to zero, gives

$$S(I + i') + Gi' - R(i - i') = 0$$

where S is the resistance of the voltmeter. For no current through the galvanometer this equation becomes

$$SI = R'i$$

where R' is the value of R that gives the balance.

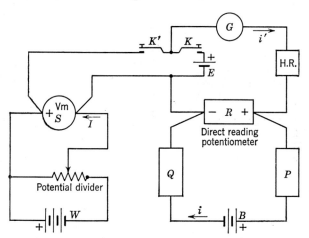

FIG. 9-9. Calibration of the voltmeter Vm.

The value of the constant current i is determined by using a standard cell of known emf. The voltmeter, with its battery, is removed from the galvanometer circuit by leaving K' open, and the standard cell is inserted in the same circuit by using the key K as shown in Fig. 9-9. The resistance in R is now adjusted to a new value R'' to give no current through the galvanometer. Then

$$E = R''i$$

since by keeping $R + P$ constant, the current is the same as before. Therefore

$$SI = E\frac{R'}{R''}$$

and this is what the voltmeter should read. If the voltmeter reading is V, the correction to be applied is

$$c = E \frac{R'}{R''} - V$$

9-12. The Calibration Curve. When the values of the true voltages are plotted against the actual readings of the voltmeter, the resulting curve is so nearly a straight line that it is not very useful. Therefore the corrections c are plotted against the voltmeter readings, as shown in Fig. 9-10. This makes it possible to use a magnified scale for the ordinates, and on this scale the small corrections appear large enough to be read easily. The plotted points are connected by the dotted line to aid the eye in noting the general trend of the corrections.

Since the corrections have been determined for only a few definite points on the scale of the voltmeter, the dotted line connecting two adjacent points has a limited significance. For example, in Fig. 9-10 the correction for 8.5 volts may not be midway between the values of the corrections for 8 and 9 volts. The dotted line shows that the correction for 7 volts is not midway between those for 6 and 8 volts. The curve shows in a general way that the correction to be applied to readings of the voltmeter between 3 and 8 volts

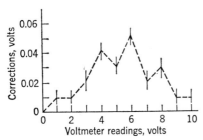

Fig. 9-10. Voltmeter calibration curve. The uncertainties in the measured values of the corrections are indicated by the lengths of the short lines.

is about +0.03 volt. Only at the points actually calibrated is the precise correction known. If corrections more accurate than these are required, they must be determined for more points, sometimes for each line on the voltmeter scale.

9-13. Measurement of Voltages Greater Than about 2 Volts. Inasmuch as most direct-reading potentiometers are designed to measure voltages up to 1.5 or 2, it is necessary to use a volt box when higher voltages are measured. The volt box is a particular form of potential divider connected to the large voltage to be measured. A definite fraction of this voltage is then measured by the potentiometer.

The connections of a volt box are shown in Fig. 9-11. Any voltage up to 230 can be applied to the total resistance of 40,000 ohms. The fall of potential over 400 ohms of this resistance can be measured with a potentiometer, and 100 times this value gives the voltage that is applied to the box.

It is evident that a small current is drawn from the source, and therefore a volt box can be used to measure only those voltages which are not altered by this current.

With the aid of such a volt box, a high-reading voltmeter can be calibrated with a low-reading potentiometer. By an increase in the resistance and the insulation, the range of a volt box can be extended to measure several thousand volts.

9-14. Comparison of Resistances by the Potentiometer. One of the accurate methods of comparing two resistances, particularly when these are not very large, makes use of a potentiometer. The two resistances to be compared are joined together in series with a battery and sufficient other resistance to ensure a steady current. This current should not be large enough to change the resistances by heating them. Other things being equal, it is desirable to have the fall of potential over each resistance about 1 volt. Let the two resistances be denoted by T and S; then the fall of potential over each will be TI and SI, respectively.

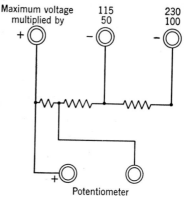

Fig. 9-11. Diagram of a volt box.

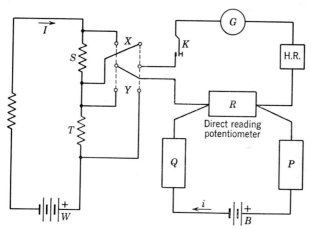

FIG. 9-12. Comparison of the resistances S and T.

The actual measurements are very simple. With the potentiometer set up as used for the comparison of emfs, measure the fall of potential over each resistance. Let the readings on the potentiometer be R' and

R'', respectively. Then from the conditions of balance,

$$TI = R'i \quad \text{and} \quad SI = R''i$$

where i denotes the current through the potentiometer resistances.

From this it follows at once that

$$T = S \frac{R'}{R''}$$

and this relation can be determined as accurately as the potentiometer measurements can be made.

9-15. Measurement of Current by Means of a Standard Cell. The potentiometer method is convenient and accurate for the measurement of current. It depends only on the measurements of a resistance and a difference of potential, both of which can be determined with a very high degree of precision. In this method the current I is passed through a four-terminal resistance R of 1 ohm or less, the value of which is known accurately. The resulting fall of potential RI is then measured with the potentiometer in terms of a standard cell, the emf of which is expressed in terms of the Weston normal cell, taken as 1.01830 volts at 20°C. Since this value was determined from an exhaustive series of comparisons with the current balance, the measurement of a current in terms of a standard cell amounts to a measurement in terms of the absolute current balance.

In the method described below, the measurements are expressed in terms of a standard cell by the regular potentiometer method.

9-16. Standard Resistance for Carrying Current. When a low resistance is used, the question of contact resistance at the connections becomes important and such a resistance is usually provided with potential terminals. This arrangement is shown diagrammatically in Fig. 9-13. MN is a low resistance with potential terminals permanently attached at A and B. When a current I is flowing through this resistance, the fall of potential between m and n is unaffected by any contact resistance at M or N, and if very little or no current is taken through m and n, resistance at these points has little or no effect on the measured fall of potential.

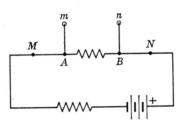

Fig. 9-13. Diagram showing the four-terminal resistance AB connected into the circuit.

It is evident, then, that the only useful resistance in this arrangement is that of the main circuit between A and B, where the potential terminals are connected, and it is the resistance of this portion that is marked on standard resistances of this type. The potential terminals are clearly seen in Fig. 9-14.

These four-terminal resistances cannot be added in series or in parallel. They can be measured and used with a potentiometer or with the Kelvin bridge.

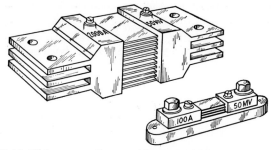

FIG. 9-14. High-current shunt. (*Courtesy of Weston Instruments.*)

9-17. Calibration of an Ammeter. *Potentiometer Method.* For the calibration of an ammeter, the instrument is joined in series with a standard resistance. The same current must therefore pass through them both. Its amount is determined by measuring the fall of potential over the standard resistance and comparing this value with the reading of the ammeter. The difference is the ammeter correction. The arrangement is shown in Fig. 9-15.

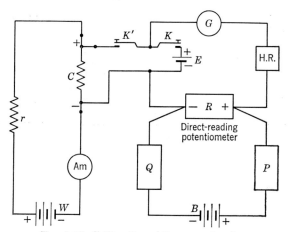

FIG. 9-15. Calibration of the ammeter *Am.*

The ammeter is connected in series with a storage battery, a variable rheostat, and the known low resistance *C*, which should be a standard manganin resistance provided with permanent potential terminals. The fall of potential over the latter is measured by the potentiometer in the usual way. The best results are obtained when this is about 1 volt, although the heat produced in the standard resistance CI^2 should be kept below 1 watt. By means of the keys K' and K, either C or the standard cell E may be used in the galvanometer circuit. In the latter case, when a balance has been obtained by adjusting R until there is no deflection of the galvanometer,

$$E = R'i$$

When the resistance C carrying a current I has been substituted for the standard cell and a balance has been obtained by readjusting R to some new value R'', we have

$$CI = R''i$$

from which
$$I = \frac{E}{C}\frac{R''}{R'}$$

This value for the current is compared with the ammeter reading I_A and the corresponding correction is

$$c = I - I_A$$

A calibration curve should be drawn, using the observed ammeter readings as abscissas and the corresponding corrections for ordinates.

PROBLEMS

9-1. Assume that the minimum current which can be detected by a galvanometer is of the order of the figure of merit. Compute the uncertainty in determining the emf of a 1-volt cell by the potentiometer method if the resistance in the galvanometer circuit is 10^3 and the figure of merit is 10^{-8} amp/mm.

9-2. A standard cell has an internal resistance of 500 ohms and an emf of 1.01830 volts. What is its terminal voltage when a current of 0.1 ma is drawn from this cell?

9-3. A saturated Weston cell is used at a temperature of 30°C. What is the emf of the cell?

9-4. An unsaturated Weston cell is used at a temperature of 30°C. What is the emf of the cell?

CHAPTER 10

MEASUREMENT OF POWER

10-1. The Measurement of Electrical Power. This usually resolves itself into the simultaneous measurement of voltage and current. The unit of power is the watt and is the power expended by a current of one ampere under a potential difference of one volt. Section 3-20 showed a simple method for measuring the power expended in a circuit by the current from a battery, with an ammeter and a voltmeter used. A single instrument combining in itself the functions of both an ammeter and a voltmeter is called a "wattmeter." With such an instrument the power may be read directly from a single scale, in the same way that the current is read from the scale of an ammeter.

10-2. The Electrodynamometer. The electrodynamometer is an instrument for measuring currents. It consists essentially of two vertical coils, one fixed in place and the other free to turn about the vertical axis common to both coils. Sometimes the movable coil is outside the other, as in the Siemens type; in other forms the movable coil is within the fixed coil. In either case, when a current flows through the movable coil, it tends to turn in the same manner as the coil of a D'Arsonval galvanometer. In the electrodynamometer the magnetic field is not due to a permanent steel magnet but is produced by the current flowing in the fixed coil. Thus the deflection depends upon the current I in the fixed coil as well as upon the current i in the movable coil, and the resulting deflection is given by

$$iI = A^2 D$$

where A^2 is a constant including all the factors relating to the size and form of the coils, etc., and also including the restoring couple of the suspension.

If the same current flows through both coils in series,

$$I^2 = A^2 D$$

and
$$I = A \sqrt{D}$$

In the Siemens electrodynamometer the coil is brought back to its initial position by the torsion of a helical spring. D is the number of divisions of the scale that measures the amount of this torsion.

155

When a coil carrying a current is suspended in a magnetic field, *e.g.*, the earth's field, it tends to turn so as to add its magnetic field to the other. If the electrodynamometer is set in such a position that the earth's field is added to the field due to the current in the fixed coil of the instrument, the deflection will be increased by a corresponding amount. If the two fields were opposed to each other, the deflection would be lessened. This effect can be eliminated by turning the instrument so that the plane of the movable coil is east and west.

Fig. 10-1. Electrodynamometer movement. (*Courtesy of Sensitive Research Instrument Corp.*)

10-3. The Use of an Electrodynamometer for the Measurement of Power. An ammeter and a voltmeter connected as shown in Fig. 10-2*a* for the measurement of a resistance will give at the same time the power expended in R and Am. Let B denote the source of the current. The voltmeter Vm measures the fall of potential E between the terminals, while the ammeter Am gives the value of the current. The product $EI = W$ gives the power in watts.

This result can be expressed in a different form. If in place of a direct-reading voltmeter there had been a large resistance of S ohms in series with a milliammeter for measuring the current i through it (Fig. 10-2*b*), then

$$E = Si \quad \text{and} \quad W = SiI$$

In this form it is seen that the measurement of power implies the product of two currents, and in Sec. 10-2 it was seen that an electrodynamometer is an instrument for measuring the product of two currents. Therefore an electrodynamometer can be used as a wattmeter if it is connected into the circuit in the proper manner for this purpose.

Let R (Fig. 10-2c) be the circuit in which the power is to be measured. The low-resistance coil a of the wattmeter W is connected in series with R as was the ammeter of Fig. 10-2a. The other coil v is joined in series with a resistance of several hundred ohms to form a shunt circuit of high resistance, and this is connected in the place of the voltmeter to measure the fall of potential over R and a. Let i denote the value of the current through this shunt circuit and S its resistance. The fall of potential is then Si. This current i through one coil of the instrument, together with

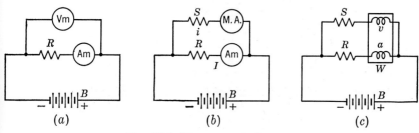

FIG. 10-2. Measurement of power.

the main current I through the other coil, will produce a deflection D proportional to the product of the two currents. From the equation of the electrodynamometer,

$$iI = A^2D$$

where A is the same constant that was previously determined.

Since the power being expended in R is $W = SiI$, we now have

$$W = SiI = SA^2D$$

If the constant of the instrument is known, then SA^2 becomes the factor for reducing the scale readings to watts. In case this factor is unity, as it can be made by adjusting the value of S, the wattmeter is said to be "direct reading."

It may be that the value of A is not known but that instead there is a calibration curve for the instrument when it is used as an electrodynamometer. In this case the value of A^2D can be obtained directly from the curve, for it is the square of the current I' that would produce the same deflection D.

Thus the power expended in R is

$$W = SI'^2$$

where S is the resistance of the shunt circuit and I' is not any real current but the current that gave the same deflection when the instrument was used as an electrodynamometer. Its value can be obtained from the calibration curve.

10-4. The Portable Wattmeter. The portable wattmeter is essentially a moving-coil electrodynamometer. The fixed coil that carries the main current is wound in two sections on a long cylindrical tube. Within this coil and midway between the two sections is the movable coil, which is wound with fine wire upon a short section of a cylindrical tube of somewhat smaller diameter than the fixed coil. In series with this movable coil is a high resistance of fine wire wound on a flat sheet of insulation to make it nearly noninductive. This coil and the high resistance serve as a voltmeter to measure the voltage factor of the power. The movable coil is supported on pivots so that it can turn about its vertical diameter. Attached to the movable coil is a long, light pointer that moves over a graduated scale.

In the position of rest, the axis of the movable coil makes an angle of about 45° with the axis of the fixed coil. When deflected so that the pointer is at the middle of the scale, the two coils are at right angles. At the extreme end of the scale the coil stands at 45° on the other side of the symmetrical position. This gives a fairly uniform scale over its entire length. A spiral spring brings the coil to the zero position and provides the torque necessary to balance the electrodynamic couple due to the currents in the two coils.

In addition to the main fixed coil there is another fixed coil of fine wire having the same number of turns as the other, so that a current sent through one coil and back through the other will produce no magnetic field at the place of the movable coil. It is then possible to compensate for the effect of the shunt current passing through the series coil, for the shunt current can be led back through this second coil and thus be made to neutralize its action upon the movable coil. When it is desired not to use the compensation coil, it is replaced by an equal resistance, this connection being brought out to a third binding post.

In the wattmeter reading up to 150 watts, the resistance of the series coil is 0.3 ohm and that of the shunt circuit is 2,600 ohms. The compensation winding is about 3 ohms.

10-5. Comparison of a Wattmeter with an Ammeter and a Voltmeter. The reading of a wattmeter can be compared with the power measured by an ammeter and a voltmeter, provided that the latter instruments are connected to measure precisely the same power as the wattmeter. This means, for the uncompensated wattmeter, that the current through the series coil of the wattmeter must be measured by the ammeter and the voltmeter must be connected so as to measure the same fall of potential as the shunt coil of the wattmeter.

This is accomplished by the connections shown in Fig. 10-3. The power thus measured is not that expended in R along but also includes the power expended in the voltmeter and in the shunt circuit of the watt-meter. Since both instruments measure this same power, the reading of the wattmeter should agree with the product of the readings from the ammeter and voltmeter.

If the wattmeter is compensated so that the power measured by it does not include the power expended in its own shunt circuit, then the ammeter must be connected so as not to measure this shunt current. As it is not possible to connect the ammeter and voltmeter so as not to measure the power expended in one or the other of them, the best arrangement will be

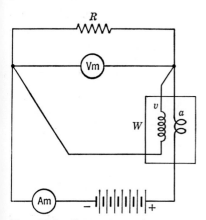

Fig. 10-3. Comparison of a watt-meter W with an ammeter and a voltmeter.

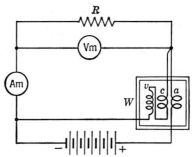

Fig. 10-4. Connections for a compensated wattmeter.

to join the voltmeter in parallel with R, as shown in Fig. 10-4. The power measured by the ammeter and the voltmeter will be that expended in both R and the voltmeter; that measured by the wattmeter will be greater than this by the amount of power expended in the ammeter. The latter can be computed from the formula rI^2 and then added to the product of the readings from the ammeter and voltmeter. With this slight correction the wattmeter reading should equal VI.

10-6. Power Expended in a Rheostat. *When Carrying a Constant Current.* The object of this exercise is to give the student some personal experience in the measurement of power and in the careful use of a variable rheostat. In this first part a variable resistance is joined in series with a large emf and considerable other resistance, so that the variations in r will not materially change the value of the current through the circuit. If desired, this variable resistance may include an ammeter, and the current may be kept always at the same value.

The wattmeter is connected to r as shown in Fig. 10-5, and readings of the power expended in the rheostat are taken for the entire range of the resistance. If the values of the latter are not known, they can be measured by one of the methods given previously. A curve should be drawn, using the resistances in the rheostat as abscissas and the corresponding amounts of power as ordinates. The report should contain a discussion explaining why this curve comes out with the form it has.

When under Constant Voltage. In this part of the exercise the arrangement is much the same as before, except that the high emf is replaced by a few cells of a storage battery, and r is now the only resistance in the

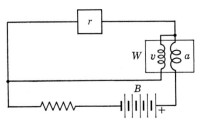

FIG. 10-5. Measurement of the power expended in r.

circuit. Starting with the largest values of r, take readings of the wattmeter and plot as before. It will not be safe to reduce r to zero, and readings should be continued only for current values that are not too large for the apparatus used. This portion of the report should give a discussion of what would probably happen if the rheostat were reduced to zero.

10-7. Measurement of Power in Terms of a Standard Cell. In the preceding chapters there have been given methods for measuring either a current or a difference of potential in terms of the emf of a standard cell. By combining two of these methods it is possible to measure power in like manner, and this is especially useful when it is desired to know accurately the value of a given amount of power. For example, in some methods of calorimetry it is necessary to have a known amount of heat supplied. Often the actual amount is not essential, but whatever it is, it must be known to a high degree of accuracy. In such cases it is convenient to generate the heat by an electric current flowing through a resistance coil and then to measure the electrical power expended in terms of a standard cell of known emf. This means the measurement of both the current and the fall of potential, as the resistance of the coil usually cannot be determined accurately under the conditions of actual use.

One convenient arrangement, which is capable of wide variation in the amount of power that can be measured, is shown in Fig. 10-6. The heating coil in which the power is expended is denoted by H. In series with this is a resistance C of sufficient current-carrying capacity not to be heated by the current through it. There is also a variable resistance r, by which the current can be brought to any desired value. The current through C is measured by the method for calibrating an ammeter, and the fall of potential over H is measured by the method for calibrating a voltmeter.

The standard cell is joined in series with a sensitive galvanometer and a high resistance. The circuit thus formed is connected to the middle of a double-throw switch S. One end of this switch is connected to B, which is a portion of a shunt around H. By adjusting A and B, the fall of potential over the latter can be made equal to the emf of the standard cell, as shown by no deflection of the galvanometer when the key K is closed. The total fall of potential over H is, then,

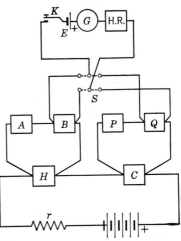

$$V = \frac{A + B}{B} E_s$$

In the same way there is a shunt circuit PQ in parallel with C, and the part Q is joined to the other end of the double-throw switch. When the switch is thrown to this side, P and Q can be adjusted to give no deflection of the galvanometer when the key is closed. This means that the fall of potential over Q has been made the same as the emf of the standard cell, and the total

Fig. 10-6. Measurement of power expended in H in terms of a standard cell E.

fall over C is computed as shown above for H. This, divided by the resistance C, gives the current through C as

$$I = \frac{E_s}{C} \frac{P + Q}{Q}$$

A little consideration will show that the main current through the battery is larger than I by the amount of current that flows through the shunt circuit PQ, and the current through H is smaller than the main current by the amount of current that flows through the shunt AB. If Q is set equal to B, then, since the fall of potential over each is the same, the currents through these two shunts will be equal. Therefore the current through H will be equal to the current through C. The power expended in H is, then,

$$W = VI = \frac{E_s^2}{B^2} \frac{(A + B)(P + Q)}{C}$$

when the two shunt currents have thus been made equal.

10-8. Calibration of a Noncompensated Wattmeter. The wattmeter in this case may be an electrodynamometer with two separate coils, or it may be the regular portable wattmeter used without the compensation coil. Its series coil (see Fig. 10-7) is connected in series with a resistance H in which can be expended the power measured by the wattmeter. There is also in series a standard resistance C whose value is known accurately and which has sufficient current-carrying capacity not to be heated by the currents used in the calibration. A variable resistance r and a storage battery z complete the main circuit. The shunt coil ab of the wattmeter is connected in parallel with H. Two accurate resistances

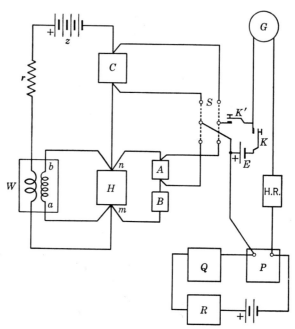

Fig. 10-7. Calibration of the wattmeter W.

A and B are also connected in parallel with H. The power measured by the wattmeter is then the total power expended in H, the wattmeter shunt circuit, and the circuit consisting of A and B. This power is the product of the current through these three in parallel and the potential difference between m and n. Each of these quantities is determined by the potentiometer.

The potentiometer is represented by the three resistances P, Q, and R. Across P the galvanometer and standard cell are joined in the usual way. When P has been adjusted for a balance

$$E = iP \qquad \text{or} \qquad i = \frac{E}{P}$$

where E is the emf of the standard cell and i is the steady constant current that is flowing through the main circuit of the potentiometer.

To determine the value of the current through the standard resistance, wires are brought from the terminals of C to the double-throw switch S. When this switch is thrown up and K' is closed, C is connected into the galvanometer circuit in the place of the standard cell. Readjusting the potentiometer for a new balance P' gives

$$IC = iP' = \frac{EP'}{P}$$

and hence

$$I = \frac{EP'}{CP}$$

To determine the fall of potential over H, the double-throw switch is thrown down, thus connecting the resistance A into the galvanometer circuit. Let i' denote the value of the small shunt current flowing through A and B. Then, adjusting the potentiometer for a balance,

$$i'A = iP'' = \frac{EP''}{P}$$

where P'' is the new reading of the potentiometer. The fall of potential over both A and B is $(A + B)/A$ times as large, or

$$V = \frac{EP''}{P} \frac{A + B}{A}$$

and this is the same as the fall of potential over H.

The power measured by the wattmeter is, then,

$$W' = VI = \frac{E^2}{P^2} \frac{P'P''}{C} \frac{A + B}{A}$$

If the reading of the wattmeter is W, the correction to be added to this reading is

$$c = W' - W$$

Different readings of the wattmeter are secured by changing the current through H. A calibration curve can be plotted, with the readings of the wattmeter used as abscissas and the corresponding corrections used for ordinates.

10-9. Calibration of a Compensated Wattmeter. For this purpose a potentiometer may be used, as in the preceding method, but as this instrument is not always at hand, a method using resistance boxes is given here. The principal difference between the compensated wattmeter and the uncompensated one is that the former measures only the power expended in the circuit to which it is attached while the latter measures, in addition to this, the power expended in its own shunt circuit.

Thus let W (Fig. 10-8) represent the wattmeter connected in the circuit to measure the power expended in H and C together. C is a standard resistance for use in the measurement of the current, and H is sufficient other resistance to give the required amount of power. As the power expended in the shunt coil is not measured by the wattmeter, it should not be measured by the standard cell; therefore C is placed inside, next to H.

The calibration circuit consists of four resistance boxes P, Q, R, and S, joined in series with a battery of a few volts sufficient to maintain a small

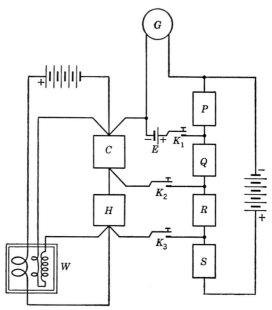

FIG. 10-8. Calibration of the wattmeter W.

constant current through the circuit. This circuit is joined through the galvanometer to the standard cell and to the wattmeter circuit in three places, as shown, each connection being provided with a key. When they are finally balanced, no current flows through any of these connections.

The measurements are made as follows: First, P is set at some convenient value, say 1,000 ohms, and some of the remaining resistances are then changed until there is no deflection of the galvanometer upon closing the key K_1. This fixes the total resistance of this circuit and it is thereafter kept constant at this amount. The fall of potential over C should be a little larger than the emf of the standard cell. A little more resistance added to P will then be required to give no deflection when the key K_2 is used. This resistance is added by varying Q and S, with their sum kept constant, until there is no deflection of the galvanometer upon closing the key K_2. This balance measures the value of the current through C, because

$$IC = i(P + Q) = (P + Q)\frac{E}{P}$$

and therefore

$$I = \frac{E}{C}\frac{P + Q}{P}$$

This is the effective current actuating the wattmeter.

The fall of potential over CH is measured in the same way. This will be greater than that over C alone, and therefore for a balance it will require a resistance greater than $P + Q$. The resistance R is now varied, with $R + S$ kept constant, until there is no deflection of the galvanometer upon closing K_3. The fall of potential over CH is then the same as that over the three resistances P, Q, and R. That is,

$$V = i(P + Q + R) = (P + Q + R)\frac{E}{P}$$

The power measured by the wattmeter is, then,

$$W' = VI = \frac{E^2(P + Q)(P + Q + R)}{P^2C}$$

If the reading of the wattmeter is W, the correction to be added to this reading is

$$c = W' - W$$

In case V is too large to be measured directly as shown here, a shunt AB may be placed around H as shown in the preceding method, and the potential fall over A along may be measured. The total fall of potential over H is then computed. The addition of this shunt will make no difference in the wattmeter, since H with its shunt now replaces H alone, and the wattmeter measures whatever power is expended in either arrangement.

10-10. Calibration of a High-reading Wattmeter. A high-reading wattmeter is one that measures large amounts of power. In calibrating such a wattmeter it is often impossible, and usually inconvenient, to expend sufficient power to bring the reading up to the high values indicated on the scale. But this is not necessary, for all that is required is that there shall be a large current through the series coil and a small current at high voltage through the shunt circuit. By using different batteries to supply these two currents, there need be no great expenditure of energy. As shown in Fig. 10-9, the battery B'', consisting of one or two large storage-battery cells, supplies a large current through the series coil of the wattmeter and the low standard resistance C. The latter should be of such a value that the fall of potential over it will be 1 volt or less, in order that this may be measured readily by the potentiometer. Since the resistance of the shunt circuit is large, it will require a large number of cells in the battery B', but these cells can be small, as only a small current will be needed. In parallel with this circuit is placed another high-resistance circuit AB, so divided that the fall of potential

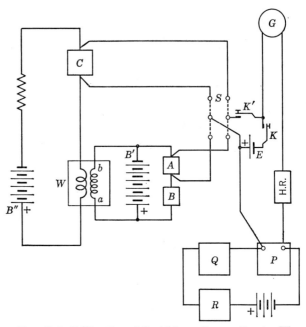

FIG. 10-9. Calibration of the high-reading wattmeter W.

over the portion A shall be about 1 volt. The calibration is then the same as given above for the case of an uncompensated wattmeter, and the wattmeter should read the product of the current through C and the voltage across A and B.

PROBLEMS

10-1. Three resistances are connected in parallel. Their values are 500, 100, and 10 ohms, respectively. If 2 watts are expended in the 100-ohm resistance, what must be the power ratings of the other resistances?

10-2. Resistances of 500, 100, and 10 ohms are connected in series. The power expended in the 500-ohm resistance is 20 watts. How much power is expended in each of the other resistances?

CHAPTER 11

MAGNETIC EFFECTS OF ELECTRICAL CURRENTS

11-1. Introduction. When iron filings are brought near a wire carrying a direct current, they tend to align themselves in a particular pattern. This alignment is brought about by the magnetic effects of the current in the wire, and these effects disappear when the current is stopped. If a cylindrical piece of steel is placed inside a coil which carries a current, the aligning force may be considerably larger than that produced by the coil itself, and when the current is stopped, an aligning force will continue to exist in the region of the steel. Under these circumstances the steel is said to be permanently magnetized and is called a magnet. Further tests with iron filings show that the magnetic effects of the permanent magnet are strongest at two regions, one near each end of the bar. These two regions are called the poles of the magnet. If the magnetized steel is now suspended from a fine thread, it is found that the magnet will always be aligned in a general north-south direction with a specific end, or pole, pointing toward the north. This north-seeking end may be labeled the north pole of the magnet while the other end may be marked south. Such a magnet (possibly in the form of a compass needle) can be used to investigate certain magnetic properties.

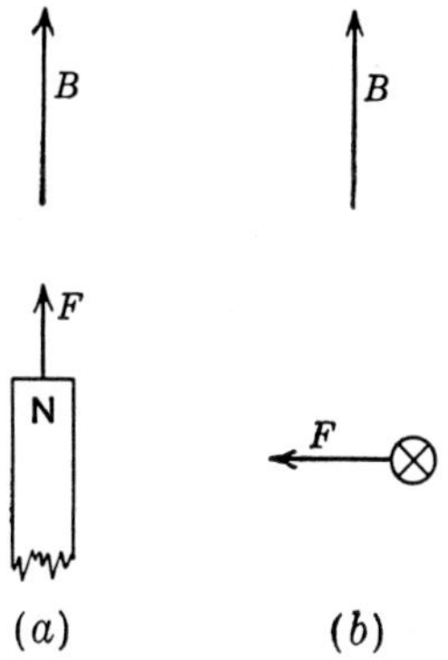

11-2. Magnetic Fields. A magnetic field exists at a point if a force is exerted on a magnetic pole when placed at this point. *The direction of the magnetic field is taken as the direction of the force acting on a north pole of a magnet.* The strength of the field is proportional to the force on a given magnetic pole.

(a)   (b)

Fig. 11-1. (a) The magnetic force acting on a north magnetic pole is in the direction of the flux-density vector B. (b) The wire is carrying *electron* current into the page. The force is at right angles to the vector B.

Quantitative definition of field properties can be given in terms of the interaction of such a field with a current-carrying wire rather than with magnetic poles. A magnetic force will be exerted on a current-carrying wire when placed at a given point in a magnetic field. The magnitude of this force depends upon the length of wire, the current flowing in the wire, and also the direction of the current with respect to the direction of the field. It is found that the force acting on the wire is not in the same direction as the force on a north magnetic pole at a given point in the field but is, rather, at right angles to the direction of the force on the pole (Fig. 11-1). The current-carrying wire is urged to move across the field. As the direction of the wire is changed, the magnitude of the force also changes, becoming zero when the wire is aligned in the direction of the magnetic field.

11-3. Magnetic Flux. It is convenient to represent the magnetic field by magnetic lines of flux. The direction of the lines of flux at a given point is in the direction of the magnetic field. The density, or closeness, of the lines represents the strength of the field. In the mks system the flux density B is expressed as webers per square meter. One weber per square meter is represented by one line of flux per square meter normal to the flux. The total flux passing through an area is then

$$\phi = \int_{area} B \, dA \qquad \text{webers}$$

or, if the field is uniform over the area considered, $\phi = BA$ webers. The flux density B can be defined in terms of the force exerted upon a current-carrying wire as follows:

$$B = \frac{F}{IL \sin \theta} \tag{11-1}$$

where B = webers per square meter
F = force exerted upon wire, newtons
I = current, amp, flowing in wire of length L
θ = angle between direction of magnetic field and current

The unit of flux density in the electromagnetic system of units is often used and can be defined from a similar equation. This unit is called the gauss and is found from

$$B' = \frac{10F'}{IL' \sin \theta} \tag{11-2}$$

B' is in gauss, F' in dynes, I in amperes, L' in centimeters. Since 1 newton = 10^5 dynes and 100 cm = 1 m, it is evident that

$$1 \text{ weber/sq m} = 10^4 \text{ gauss}$$

The unit of flux in the emu system is the maxwell, and a flux density of one gauss is equivalent to one line of flux per square centimeter or one maxwell per square centimeter.

11-4. Magnetic Effect of a Current-carrying Wire (Ampère's Law). An electrical current always has associated with it a magnetic field. It is the interaction of the external field with the field of the current-carrying wire which forms the basis for the definition of the flux density B. Figure 11-2 represents a wire carrying an electron current I in the direction indicated. dB is that part of the flux density at the point P produced by a part of the wire dl. According to Ampère's law, the magnitude of

$$dB = \frac{KI\,dl\,\sin\alpha}{r^2} \tag{11-3}$$

If dB is expressed in webers per square meter, I in amperes, dl in meters and r in meters, the constant K is $\mu_0/4\pi$ in vacuum. μ_0 is the permeability

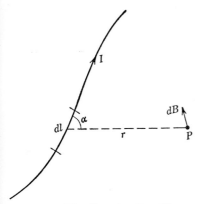

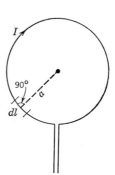

Fig. 11-2. The flux density dB produced by the element of length dl carrying an electron current I is perpendicular to and directed out of the page.

Fig. 11-3. Flux density at the center of a circular loop of radius a. The angle between a and dl is always 90°.

of empty space and has a value of $4\pi \times 10^{-7}$ weber/amp-m. Thus the numerical value of K in the mks system is 10^{-7}.

(If dB is in gauss, I in amperes, dl in centimeters, and r in centimeters, the numerical value of K is $\frac{1}{10}$.)

The direction of dB at the point P is perpendicular to both r and dl and is therefore normal to the page in Fig. 11-2. One further point needs to be specified in order to describe completely the vector dB, namely, its sense (*i.e.*, whether it is into or out of the page). This can best be remembered by imagining that the wire is grasped in the left hand with the thumb pointing in the direction of flow of electron current. The fingers will then point in the direction of the vector dB.

The relationships discussed above allow one to evaluate the total flux density B at a point by summing the contributions of all parts of the wire vectorially.

Example. Find the flux density at the center of a circular loop of wire of radius a m. The current in the wire is I amp (Fig. 11-3).

From Ampère's law dB at the center of the loop arising from the portion of the wire dl is

$$dB = \frac{10^{-7}I \, dl}{a^2}$$

The total flux density at the center of the loop arises from the contribution of all the individual segments and is therefore

$$B = \frac{10^{-7}I}{a^2} \int_0^{2\pi a} dl = \frac{2\pi \times 10^{-7}I}{a} \qquad \text{webers/sq m}$$

The field is perpendicular to the page and directed outward.

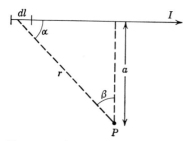

FIG. 11-4. Flux density at a distance a from a long, straight wire. The electron current is I.

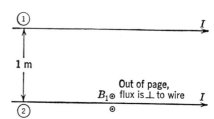

FIG. 11-5. Force between two parallel wires 1 m apart. The current in each wire is 1 amp if the wires experience a force of 2×10^{-7} newton per meter of length.

As a second example, consider the flux density at a distance a from a long, straight wire carrying a current I (Fig. 11-4). Again

$$dB = \frac{10^{-7}I \, dl}{r^2} \sin \alpha \qquad \text{and} \qquad B = 10^{-7}I \int_{-\infty}^{+\infty} \frac{\sin \alpha \, dl}{r^2}$$

Since $l = a \tan \beta$, $dl = a \sec^2 \beta \, d\beta$, $\sin \alpha = \cos \beta$, and $r = a/\cos \beta$, we can write

$$B = 10^{-7}I \int_{\beta=-\pi/2}^{\beta=+\pi/2} \frac{1}{a} \cos \beta \, d\beta = \frac{2 \times 10^{-7}I}{a}$$

It is this relationship, combined with the definition of B, which forms the basis for the definition of the ampere of current; i.e., 1 amp is the current in each of two long, parallel wires placed 1 m apart in vacuum when the force acting on each wire is 2×10^{-7} newton/m. In Fig. 11-5, the flux density at wire 2 produced by the current in wire 1 is B_1. The force on wire 2 is thus

$$F_2 = B_1 I \, dl \sin \theta = \left(\frac{2 \times 10^{-7}I}{1} \right) I(1)(1)$$

If the force F_2 is 2×10^{-7} newton, I has the value of 1 amp.

11-5. Flux Density within a Solenoid. The flux density on the axis of a circular loop of wire carrying a current I (Fig. 11-6a) is

$$B = \frac{\mu_0}{4\pi} \frac{(2\pi a^2)I}{(a^2 + x^2)^{3/2}}$$

The radius of the loop is a m; x represents the distance from the center of the loop to the point at which B is measured. The flux is directed along the axis. (Proof of this is left as an exercise for the student.)

A solenoid can be considered as a succession of such loops placed next to each other (Fig. 11-6b), there being n loops or turns per meter of length. If the total length of the solenoid is l m, the flux density at the center can be written

$$\begin{aligned}
B &= \frac{\mu_0}{4\pi} (2\pi a^2)I \int_{-l/2}^{l/2} \frac{n \, dx}{(a^2 + x^2)^{3/2}} \\
&= \frac{\mu_0}{4\pi} (2\pi a^2)nI \left. \frac{x}{a^2(a^2 + x^2)^{1/2}} \right|_{-l/2}^{l/2} \\
&= \frac{\mu_0}{2} nI \frac{l}{[a^2 + (l^2/4)]^{1/2}} \qquad \text{weber/sq m}
\end{aligned}$$

For a long solenoid ($l \gg a$) the field is uniform throughout and equal to $\mu_0 nI$ weber/sq m. ($B = 0.4\pi nI$ if n is expressed as turns per centimeter, I is in amperes while B is in gauss.)

11-6. Direction of the Force on a Current-carrying Wire in an External Magnetic Field. When a current-carrying wire is placed in an external magnetic field, the field in the neighborhood of the wire is changed and at each point it is the resultant of these two superimposed fields. A qualitative inspection will show that the total field on one side of the wire is increased (since both the external field and the field of the current in the wire are in the same direction) while on the oppo-

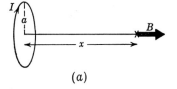

(a)

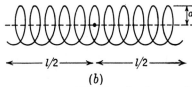

(b)

Fig. 11-6. (a) The flux density on the axis of a loop is directed along the axis. (b) The flux density at the center of a solenoid can be obtained by considering the influence of a large number of loops uniformly distributed from $-l/2$ to $+l/2$.

site side of the wire the fields oppose each other. *The force on the wire is always in the direction such as to tend to move it from the stronger toward the weaker field.* This is a direct consequence of the law of conservation of energy and will be discussed in greater detail in the treatment of induced emf.

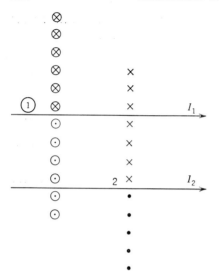

FIG. 11-7. The fields from the wires reinforce above the upper wire and below the lower wire but oppose each other between the wires. They will experience a force of attraction.

To illustrate this principle, consider two parallel wires carrying currents in the same direction (Fig. 11-7). The direction of the lines of flux for wire 1 are illustrated by dots inside circles to represent flux coming out of the page and xs inside circles to represent flux going into the page. The dots and xs of wire 2 are not circled. It is seen that the two fields are in opposition in the region between the wires but are in the same direction above and below the wires. The force will thus tend to push the wires toward each other.

11-7. Torque Acting upon a "Current Loop" Suspended in a Magnetic Field. A rectangular coil $ABCD$ (Fig. 11-8) is suspended in a uniform magnetic field of flux density B webers/sq m. An electron current of I amp flows in a clockwise direction. The force acting on the side AB is F_{AB} and is directed into the page, $F_{AB} = BIh$ newtons. The torque produced by this force is $\tau = BIh[(w/2)\cos\gamma]$. ($\gamma$ is the angle between the direction of B and the plane of the coil.) A similar torque of equal magnitude and in the same sense is produced by the force on side CD. Thus the total torque is $\tau = BI(hw)\cos\gamma$ newton-m. The quantity hw is the area of the coil. (The force on BC is equal and opposite to the force on DA and produces no torque about the axis indicated.)

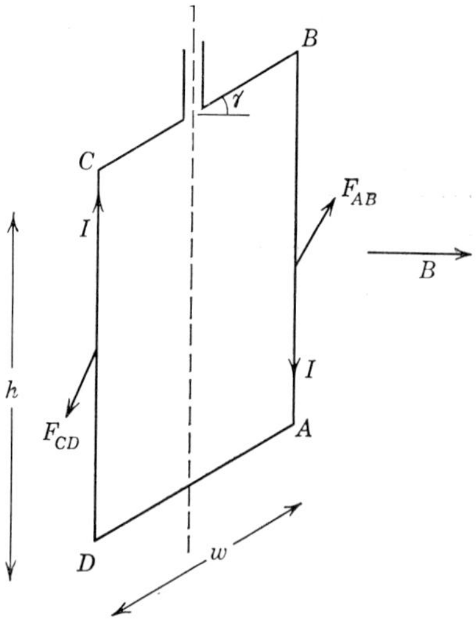

In a similar manner, the torque on a circular loop can be derived (Fig. 11-9). The plane of the circle makes an angle γ with the direction of B. The radius of the loop is r m. The force acting on the vertical component of a current element dl is into the page and equal to $BI\,dl\cos\eta$

FIG. 11-8. The torque acting on a rectangular coil is $BIA\cos\gamma$.

newtons. (η is the angle between the horizontal and r.) The torque produced by this force about a vertical axis through the center is

$$\tau = BI\, dl \cos \eta \, (r \cos \eta \cos \gamma)$$

Since $dl = r\, d\eta$, we can write $\tau = BIr^2 \cos \gamma \cos^2 \eta \, d\eta$, and the total torque acting on the coil is

$$\tau = BIr^2 \cos \gamma \int_0^{2\pi} \cos^2 \eta \, d\eta$$

$$= BIr^2 \cos \gamma \left(\frac{\eta}{2} + \frac{1}{2} \sin \eta \cos \eta \right) \Big|_0^{2\pi}$$

$$= BIA \cos \gamma$$

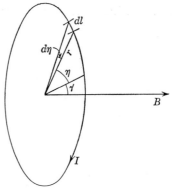

FIG. 11-9. The torque on a circular coil about a vertical axis through its center is $BIA \cos \gamma$.

In a similar manner it can be shown that a plane coil of arbitrary shape carrying a current of I amp is acted on by a torque $\tau = BIA \cos \gamma$ if the plane of the coil makes an angle γ with the direction of B. The quantity IA is called the *magnetic moment* or dipole moment of the coil.

11-8. Alignment of Current Loops in Magnetic Fields. When a current loop which is free to rotate is placed in an external magnetic field, it will turn under the influence of the magnetic torque until the angle between the plane of the coil and the direction of the magnetic field is 90° (Fig. 11-10). At this position the field of the current loop has the same direction as the external field. This is a position of stable equilibrium, since any displacement of the coil will result in the production of a torque which tends to return it to this position.

According to the principle outlined, one should expect that several small current loops which are free to rotate could align themselves in the manner shown in Fig. 11-11 owing to the mutual action of their magnetic fields. This results in an over-all strengthening of the field in the immediate region of the loops but produces essentially no magnetic effects elsewhere. The lines of flux are much like those produced by a minute toroidal winding.

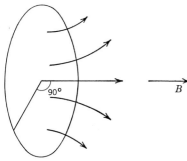

FIG. 11-10. When a current loop having a magnetic moment IA is placed in a magnetic field, it will come to equilibrium with its own field in the direction of the aligning field B.

Let us next suppose that a strong, external magnetic field B_0 is applied to this assembly of coils shown in Fig. 11-11. If the external field is sufficiently strong, the torque produced by its action on the individual current loops may be sufficient to decouple them from their previous alignment and realign them as shown in Fig. 11-12. Such an alignment will clearly bring about an increase in the total flux density over a relatively large region. Within the material, the total flux density B will be that due to the external field B_0 plus that due to the internal alignment of

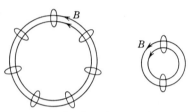

the magnetic moments B_i, so that $B = B_0 + B_i$. The quantity B_i depends both upon the properties of the material and upon the external field. That B_i depends upon the properties of the material is evident if one considers such diverse media as wood and iron. The atoms or molecules of the wood have essentially zero magnetic moment and as such cannot contribute to the field, so that the flux density $B = B_0$. In other words, the field within the wood is essentially identical with the field which existed in the region before the wood was placed

Fig. 11-11. In the absence of an external aligning field, the atoms of a paramagnetic substance will be coupled to each other and will produce an essentially random array. There will be zero magnetic effect at distances large compared with the atomic size.

there. On the other hand, iron has a large number of magnetic dipoles per unit volume which can be aligned by the external field, so that the quantity B_i for iron can be much greater than B_0.

It should be recalled that the "current loops" represent, in reality, the orbital motion of electrons about the nucleus of individual atoms. (A spinning electron also has a magnetic moment.) Since the atoms of a substance are in constant motion as a result of thermal agitation, the degree of alignment will be dependent upon the magnitude of the aligning field B_0.

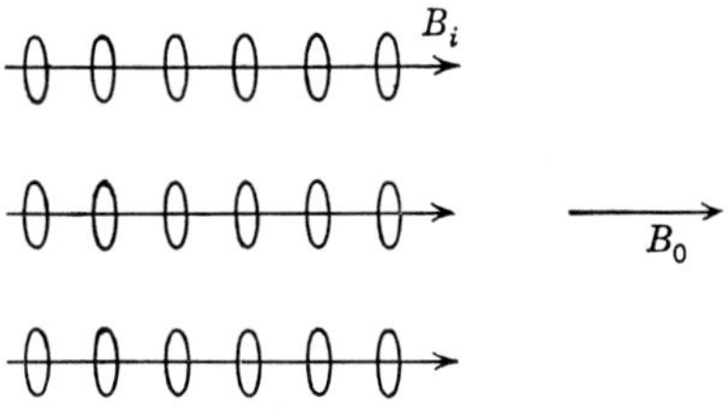

Substances which are made of atoms having such permanent magnetic moments are said to be paramagnetic and will always cause a strengthening of the magnetic field by the alignment process described above.

Fig. 11-12. Under the influence of the external field B_0, the atomic dipoles or current loops are aligned to reinforce each other as well as the external field.

Most substances have a permanent magnetic moment which is essentially zero and as such contribute nothing to the magnetic field; that is, $B_i = 0$.

A few substances exist which have zero permanent magnetic moment *by themselves.* However, in an external magnetic field B_0, a magnetic moment may be *induced* which opposes the inducing field B_0 and thus tends to produce a slight over-all weakening of the magnetic field; $B = B_0 - B_i$. Such substances are said to be diamagnetic.

11-9. The Magnetic Circuit. When a wire is wound into a helix and this solenoid carries a current, all the paths traced by the compass needle or by the iron filings are found to pass through the inside of the solenoid from one end to the other, then out through the surrounding space to the starting points. It is therefore natural to associate the idea of a magnetic *circuit* with the phenomena of magnetism.

The magnetic flux is continuous and forms a closed circuit with itself. At one place it may be spread out over a large cross section, and in another part of the circuit it may be confined within narrow limits, but the total amount of magnetic flux is the same for each cross section of the circuit.

Similar paths can be traced through the region around a permanent magnet, but in this case the compass needle cannot be carried through the magnet itself, and other methods of measurement are necessary to show that the paths are continuous through the magnetized steel. The whole sheaf of paths in any given case maps out the magnetic circuit for that case.

11-10. Magnetomotive Force. The magnetic flux in the air around a wire carrying a current is due to the stream of electrons along the wire. In fact, there is such a strict proportionality between them that a current is usually measured by means of the magnetic flux associated with it. With a solenoid it is easily seen that the number of turns of wire is as important as the value of the current. A current of n amp through *one* turn of wire gives the same magnetic effect as 1 amp through n turns of wire that occupy the same place as the single turn of wire. The magnetic effect of a current is thus measured in "ampere-turns," and every current has a circuit of at least one turn.

In the case of a steel magnet or a magnetized piece of iron, the magnetic flux is similarly due to the circulatory motion of electrons, but in this case the electrons are moving within the atoms of iron or spinning on their axes.

"Magnetomotive force" is the name that has been given to this magnetic influence of a current. It can be measured in ampere-turns or in other units proportional to ampere-turns. The effect of magnetized iron or steel can be expressed likewise in terms of the equivalent ampere-turns. There can be no magnetic flux that is not associated with the corresponding ampere-turns of electrons in a wire or within the atoms of a magnetized substance. Inasmuch as it is not possible to count the turns or to measure the current to determine the equivalent ampere-turns of a piece of magnetized steel, it is customary to speak of the effect of the amperes of current in the actual turns of wire as the magnetomotive force and to measure experimentally the relation of these ampere-turns to the total magnetic flux in the circuit due to both the current and the magnetized steel.

11-11. The Ampere-turn. The words "magnetomotive force" are often abbreviated by the initials mmf. The unit of magnetomotive force is called an "ampere-turn," and it is the mmf due to an ampere in one turn of wire.

The magnetomotive force due to a current in a coil of N turns is given by the relation

$$\text{mmf} = NI \qquad \text{amp-turns}$$

where I denotes the value of the current in amperes.

11-12. Representation of Magnetomotive Force. The turns of wire in a solenoid are counted along the length of the solenoid. Since magneto-

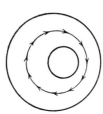

Fig. 11-13. The arrows show the amount and distribution of the mmf for a circuit of either iron or wood. The mmf depends solely upon the number of ampere-turns of the magnetizing current and is independent of the material of the magnetic circuit.

motive force is measured by the number of ampere-turns that are linked with the magnetic circuit, the mmf is measured likewise along the length of the magnetic circuit. Whether the ampere-turns are uniformly distributed along the magnetic circuit or are bunched together over a short distance, the total mmf is the same amount. This mmf is distributed along the magnetic circuit in proportion to the reluctance of the various parts of the circuit,[1] similarly to the volts along an electric circuit. This distribution can be represented by a series of marks or arrowheads properly spaced along the length of the magnetic circuit, the total number of arrowheads being made equal to the total number of ampere-turns. As shown in Fig. 11-13, this is quite different from the representation of magnetic flux, which is measured by the maxwells through the total cross section of the circuit.

11-13. Arrows per Centimeter. Where the arrows are close together in the representation of mmf (Fig. 11-14), there the magnetic intensity H is great; where they are far apart, H is less. The value of H at any point of the circuit is given by the number of arrows per centimeter at that point. This value may change abruptly where the circuit passes from one medium into another, as from

---

[1] If the ampere-turns are bunched at one place, the magnetic circuit may not be the same circuit that it was when these ampere-turns were uniformly distributed.

iron into air, although a wire and ballistic galvanometer would show that the magnetic flux is the same in each medium.

11-14. Magnetomotive Force in a Divided Circuit. The mere fact that a magnetic circuit is divided into two parallel branches does not change the number of ampere-turns that are linked with the total circuit. If the total mmf along a given divided circuit, as in Fig. 11-15, is 9 amp-turns, the mmf along each branch is 9 amp-turns also. Even if one branch is wholly in air and the other wholly in iron, the ampere-turns around one branch are the same as the ampere-turns linked with the other branch, and therefore the amount of mmf distributed along the length of each branch must be the same.

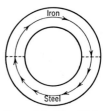

Fig. 11-14. This shows the distribution of mmf in a circuit half of steel and half of the more permeable iron.

11-15. Relations in the Magnetic Circuit. In a magnetic circuit like the ring shown in Fig. 11-15 but uniformly wound with N turns of wire instead of the three turns shown in the figure, the amount of flux ϕ depends upon a number of factors. Primarily, ϕ depends on the current I and the number of turns of wire wound on the ring. It also depends on the size of the ring, being greater in a ring of greater cross section A and inversely proportional to the length[1] l of the magnetic circuit for the same values of N and I. Experimental investigation also shows that the amount of flux depends upon the quality of the material of the ring. With a ring of iron the flux may be several hundred or even several thousand times as much as in a ring of wood or brass with all of the other factors the same. This help that is rendered by the electrons in the atoms of the material is expressed by the factor μ, which is called the "permeability" of the substance.

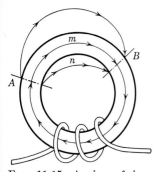

Fig. 11-15. A ring of iron magnetized by three turns of current. The arrows indicate a mmf of 9 amp-turns and show how this is distributed. Between A and B the mmf along one path is the same as along any other path.

Collecting these factors together gives

$$\phi = IN\left(\frac{A}{l}\right)\mu k \qquad (11\text{-}4)$$

where k is a proportionality factor depending upon the units in which the other quantities are measured. The factors in this expression can be rearranged in different ways to suit various purposes.

[1] The ring is supposed to be thin in the radial direction, so that the mean circumference l is practically the same as the outer or the inner circumference.

11-16. Reluctance. One way of looking at the magnetic circuit is similar to Ohm's law for the electric circuit. All the factors in Eq. (11-4) that relate to the size and material of the magnetic circuit may be collected together and written as

$$R = \frac{l}{\mu A}$$

From analogy with the resistance in an electric circuit, R is called the "reluctance" of the magnetic circuit. A circuit that does not have a constant cross section may be considered as a series of circuits having lengths l_1, l_2, l_3, etc., and cross sections A_1, A_2, A_3, etc. The total reluctance is then

$$R = \frac{l_1}{\mu_1 A_1} + \frac{l_2}{\mu_2 A_2} + \frac{l_3}{\mu_3 A_3} + \text{etc.}$$

Iron in its various forms is one of the best "conductors" of magnetic flux. Most other substances are rather poor "conductors," but there are no "insulators" of magnetic flux. Even when a magnetic circuit is made largely of iron, the magnetic flux does not follow the iron exclusively, but there is some flux that finds a path through the surrounding air. It is extremely difficult to build a circuit, even of soft iron, in which all the flux will follow the iron path. If two paths are open to the magnetic flux, it will divide just as an electron current would do between two wires in parallel. The total reluctance of a circuit includes the effect of the reluctance of such parallel leakage paths.

In iron and similar magnetic substances, there is a difference, too, between electric resistance and magnetic reluctance, inasmuch as the former is constant for all ranges of current, while the value of the reluctance of a magnetic circuit depends upon the amount of magnetic flux through it, being greater for very small or very large values of the magnetic flux than it is for moderate values.

11-17. Law of the Magnetic Circuit. In the preceding section, all factors in Eq. (11-4) relating to the size and material of the magnetic circuit were grouped together under the name "reluctance." In a similar manner, all the factors that relate to the effect of the current can be grouped together. (The constant k can be lumped into this factor.) We can then write

$$\text{Magnetic flux} = \frac{\text{mmf}}{\text{reluctance}}$$

$$\phi = \frac{kNI}{l/\mu A}$$

In the mks system of units, ϕ is in webers, I in amperes, l in meters, and A in square meters. In this system the value of k is 1. The magnetomotive force (NI) would be expressed in ampere-turns.

ϕ is in maxwells if I is in amperes, l in centimeters, and A in square centimeters. The value of k is 0.4π, and the magnetomotive force $(0.4\pi NI)$ *is given the name gilbert.*

11-18. Magnetic Intensity H. A given amount of magnetomotive force may be distributed over a magnetic circuit of considerable length, or it may be concentrated in a circuit of shorter length. In the latter case the magnetic effect is more intense than in the former case. These differences can be described by the term "magnetic intensity," which is taken as the magnetomotive force per unit of length and is denoted by the letter H.

$$H = \frac{d(\text{mmf})}{dl} \tag{11-5}$$

If the mmf is distributed uniformly we can write $H = \text{mmf}/l$.

When the flux density is measured in webers per square meter, the appropriate unit for the magnetic intensity H is ampere-turns per linear meter.

When the flux density is in gauss, H is commonly expressed as gilberts per centimeter, that is, 0.4π amp-turns/cm. This unit is commonly called the *oersted*. One oersted equals one gilbert per centimeter.

11-19. The Relation between B and H. Equation (11-4) can be rearranged to yield $\phi/A = \mu(kNI/l)$. But in any system of units ϕ/A equals the flux density B and kNI/l is the magnetic intensity H. Thus

$$B = \mu H$$

If B is in webers per square meter, H is expressed as ampere-turns per meter. The permeability is expressed in webers per ampere-meter. The value of the permeability of empty space μ_0 has the value $4\pi 10^{-7}$ weber/ amp-m.

If B is expressed in gauss, H is in oersteds and μ is expressed in gauss per oersted. The numerical value for the permeability of empty space in this system is $\mu_0 = 1$.

It should be kept in mind that B and H are evaluated at each point in a complex arrangement.

11-20. Circuit Integral of H. The magnetic intensity H is defined by Eq. (11-5). It thus follows that the integral of H around a closed path is equal to the total mmf around this path:

$$\oint H \, dl = \text{mmf}$$

In the mks system the mmf is just the number of ampere-turns cutting the area enclosed by the path. (In the gaussian system the mmf is $0.4\pi NI$.)

Example. A toroidal winding is shown in Fig. 11-16, having a total of N turns and carrying a current of I amp. Consider the integral taken around the dotted path: $H(2\pi r) = NI$ or $H = NI/2\pi r$. It is seen that the magnetic intensity decreases within the windings as r increases.

An interesting case is encountered when the radius of the toroid is allowed to increase while keeping the number of turns per unit length constant. As the radius approaches infinity, the variation across the toroid disappears and H has a value

$$H = nI$$

n is the number of turns per meter of length. This is, of course, the relationship found for a solenoid, since a toroid having an infinite radius of curvature is a solenoid.

11-21. Permeability μ. In discussing the relations in the magnetic circuit in Sec. 11-15, the factor μ was introduced to represent the influence of the medium upon the resulting magnitude of the magnetic flux. In Sec. 11-19 this same factor remains in the relation between B and H.

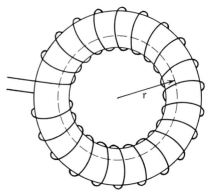

Fig. 11-16. The magnetic intensity within the solenoid is $H = NI/2\pi r$. N is the total number of turns each of which carries a current of I amp.

This factor is called the "magnetic permeability" of the medium, and even in a vacuum it is not an abstract number, because B and H do not become quantities of the same nature.

The permeability is always a positive quantity. For most substances its value is nearly 1 gauss/oersted. In a few materials it is slightly less than 1, and these substances are called "diamagnetic." Substances for which the permeability is slightly greater than 1 are called "paramagnetic." For ferromagnetic substances like iron, nickel, cobalt, etc., the permeability may reach a value of several thousands. The unit of permeability has no name. A magnetic circuit of large permeability will have a correspondingly small reluctance. The reciprocal of permeability is called "reluctivity."

11-22. Relative Permeability μ'. Another way in which the term "permeability" is sometimes used is in the comparison of the amounts of magnetic flux in two media that are subjected to the same magnetic intensity. For example, if the flux density is B_i in a certain circuit consisting of iron, and if the flux density is B_0 for the same circuit when the iron has been replaced by a vacuum, then

$$\frac{B_i}{B_0} = \mu' = \frac{\mu}{\mu_0}$$

where μ' is the relative permeability of iron as compared with a vacuum.

The relative permeability, being the ratio of two similar quantities, is an abstract number. This is not the case with the permeability μ, which is the ratio of quantities B and H that are dissimilar in nature.

11-23. Intrinsic Magnetization. The magnetic flux through a magnetized circuit may be considered as the sum of two superimposed parts— one portion due to the current in the magnetizing solenoid and a second part due to the electronic mmf within the substance.

If the magnetizing solenoid were in vacuum, there would be some flux through it, and

$$B_0 = \mu_0 H$$

where the subscripts refer to the values in vacuum. When a magnetic substance is within the solenoid, the total flux density is this B_0 *plus* the flux density B_I that is contributed by the substance. Thus

$$B = B_0 + B_I \tag{11-6}$$

When H is made larger and larger, the value of B continues to increase due to the part B_0. The part B_I seems to reach a definite limit for each substance, which is reached when all of the movable electronic orbits, or whatever parts are free to turn, are oriented as nearly as possible into alignment with the applied mmf.

11-24. Intensity of Magnetization J. A conception of this quantity is obtained from the following considerations. If a bar of uniformly magnetized steel is placed in a magnetic field of strength B, the maximum torque exerted by the field is

$$T = J \times \text{volume} \times B$$

or
$$J = \frac{T}{B \times \text{volume}}$$

Hence the intensity of magnetization J for a given piece of material is given by the maximum torque per unit volume that would be exerted by a unit magnetic field. This quantity J is called the "intensity of magnetization." It should be noted that this is not magnetic intensity H.

11-25. Susceptibility. The susceptibility κ of a substance is the ratio of the intensity of magnetization J to the value of H necessary to produce it, or

$$\kappa = \frac{J}{H}$$

There is a simple relation between κ and μ for a substance in a given magnetic condition. Substituting in Eq. (11-6), Sec. 11-23, the values of the different flux densities with the aid of the relation $B_I = \mu_0 J$ gives

$$\mu H = \mu_0 H + \mu_0 \kappa H$$

or $\qquad \mu = \mu_0 + \mu_0 \kappa \qquad$ weber/amp-m

PROBLEMS

11-1. Find the flux density on the axis of a circular coil and d m from its center. The coil has a radius of r m, N turns of wire, and the current in the coil is I amp.

11-2. A rectangular coil of 8 turns, 0.1 m high and 0.01 m wide, is suspended in a uniform magnetic field of flux density 1.5 webers/sq m. The plane of the coil makes an angle 30° with the direction of the lines of flux. Calculate the torque acting on the coil if the current is 4 amp.

11-3. Two long, parallel wires are 5 cm apart. One carries a current of 10 amp to the right; the other carries a current of 4 amp to the left. Find the magnitude and direction of the force between the wires.

11-4. Express the units of μ_0 in terms of mass, length, and time.

11-5. The horizontal component of the earth's magnetic field is 0.2 gauss. A vertical wire 10 cm long carries an electron current of 40 amp downward. Find the magnitude and direction of the force on the wire.

11-6. A particle of mass M and charge $-q$ is moving horizontally with a velocity of V m/sec in a uniform magnetic field of flux density 1.5 webers/sq m. The field is directed vertically downward. Calculate the magnetic force acting on this particle and the radius of curvature of its path.

11-7. Assume that the electron in a hydrogen atom revolves about the nucleus in a circular orbit of 0.5×10^{-8} cm radius. Calculate the magnetic flux density at the nucleus produced by the orbital motion of the electron. (The charge on the nucleus is $+1.6 \times 10^{-19}$ coulomb and of the electron is -1.6×10^{-19} coulomb. The mass of the electron is 9×10^{-28} gram.)

11-8. Considering only the orbital motion as in Prob. 11-7, what is the magnetic moment of the hydrogen atom?

11-9. Prove that the magnetic moment resulting from the revolution of a point charge is $ep/2mc$. e is the charge, p the angular momentum, m the mass of the revolving particle, and c the velocity of light.

11-10. The atom of Prob. 11-7 is placed in a magnetic field of flux density B webers/sq m with its magnetic moment aligned in the direction of B. Calculate the energy required to rotate the magnetic moment to a position antiparallel to B. (This is closely related to the normal Zeeman effect in spectroscopy.)

11-11. When an atom of magnetic moment J is placed in a uniform external magnetic field, a torque acts upon the atom to align it with the external field. When the atom is placed in a nonuniform field, an additional unbalanced force will act upon the atom, tending to move it parallel to the lines of force. Prove that this force is equal to $F = (dB/dx)J$. The gradient of the field is dB/dx. (This is the basis of the Stern-Gerlach experiment.)

11-12. A flux density of 1 weber/sq m is to be produced in the iron ring in the accompanying diagram. The relative permeability μ/μ_0 of the iron, is 500. Assume that all the flux is confined to the ring.

a. What is the reluctance of the circuit?

b. How many ampere-turns are required to produce this flux density?

c. If the windings are uniform, what fraction of the flux density is due to the current in the windings, that is, B_0/B?

d. What fraction of the flux density is due to the aligned magnetic moments, that is, B_i/B?

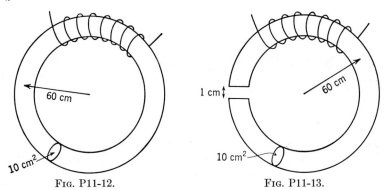

FIG. P11-12.   FIG. P11-13.

11-13. A slice of the iron 1 cm thick is removed from the ring in Prob. 11-12. Assume that the magnetic flux in the air gap is confined to an area of 10 sq cm.

    a. What is the reluctance of the air gap?

    b. What is the reluctance of the iron?

    c. How many ampere-turns are required to produce a flux density of 1 weber/sq m?

    d. What is the mmf across the gap?

    e. What is the magnetic intensity H within the gap?

    f. What is the magnetic intensity within the iron?

CHAPTER 12

FARADAY'S LAW, MUTUAL INDUCTION, SELF INDUCTION, AND MAGNETIC MEASUREMENTS

12-1. Introduction. When a wire is moved across lines of flux, it causes a deflection of the ballistic galvanometer to which it is attached. The quickness with which the wire moves makes no difference in the galvanometer deflection, but if the wire is moved across a greater area, the deflection is correspondingly increased.

If the test wire is moved in the direction of the lines of flux, there is no deflection of the galvanometer, since there has been no change in the amount of flux enclosed by the wire.

If by any means whatsoever the magnetic flux enclosed by this loop of wire is made to change, there is a corresponding deflection of the galvanometer. Magnetic flux is detected, then, by the deflection of the galvanometer, and the magnitude of this deflection may be taken as an indication of the total *change* in the flux enclosed by the wire and its connections to the galvanometer. It does not matter whether the wire is moved across the flux or the flux is moved across the wire or if the flux density changes without any motion; the effect on the galvanometer is the same.

The deflection of the galvanometer shows that a transient current has passed through it. This current is due to an emf in the galvanometer circuit. That it is an emf and not something like a condenser discharge is shown when resistance is added to the galvanometer circuit. The deflection is then decreased in proportion to the increase in the total resistance of the circuit.

Since the quickness with which the change in flux is made does not change the deflection, it is evident that when this change is made in a short time, the current, and therefore the emf, is correspondingly larger during this short time. That is, $Q = \int_0^T i \, dt = \int_0^T \frac{e}{R} \, dt = $ constant.

12-2. Faraday's Law. The effects which have been discussed above can be described quantitatively by Faraday's law:

$$e \sim \frac{d\phi}{dt} \tag{12-1}$$

The *instantaneous* voltage induced in a circuit is proportional to the rate at which lines of magnetic flux are cut by the circuit.

In the mks system

$$e = \frac{d\phi}{dt} \tag{12-2}$$

if e is in volts and $d\phi/dt$ is in webers per second.

In a commonly used mixed system of units in which e is expressed in volts and $d\phi/dt$ in maxwells per second, the constant of proportionality is 10^{-8} and the equation can be written

$$e = \frac{1}{10^8} \frac{d\phi}{dt} \tag{12-3}$$

12-3. Lenz's Law. In addition to the magnitude of the induced voltage or emf as given by Faraday's law, it is also necessary to know the direction of this voltage. Lenz's law, which is merely an application of the principle of conservation of energy, is intended to give this direction. The law can be stated as follows: *The direction of the induced emf is such as to oppose the change which causes the emf.*

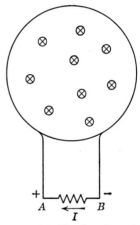

Two precautions should be noted in applying Lenz's law. First, the actual opposition to a change is brought about by the current which results from the induced emf acting on a closed circuit. If a voltage is induced in an open circuit, there will be no real opposition but the induced emf will still be present and in the same direction as for a closed circuit. Second, care must be exercised in assigning polarities to the ends of wires in which a voltage is induced. It should be remembered that these wires are the seat of the emf and as such play the same role as batteries. In a battery the electron

Fig. 12-1. The flux through the coil is increasing in the downward direction. The electron current is from B toward A.

current flows from the positive terminal to the negative terminal *within the battery.*

Example. A horizontal coil of 25 turns of wire is in a region in which the flux density is vertically downward and increasing at the rate of 0.2 weber/sq m-sec. The coil encloses an area of 0.1 sq m, and its terminals are connected to a 400-ohm resistance A-B (Fig. 12-1). Find the magnitude of the induced emf, the direction of the current, and the relative polarity of A and B.

$$e = 25(0.2)(0.1) = 0.5 \text{ volt}$$

The flux produced by the induced current must oppose the external increase of flux in the downward direction. Therefore the electron current must flow in a clockwise direction (looking downward). The magnitude of the current is $I = 0.5/400 = 0.00125$ amp, and since the electrons flow through the external resistance from B to A, B is at a lower potential than A.

Example. Using the same arrangement as in the previous example, the flux is still downward but is decreasing at the rate of 0.2 weber/ sq m-sec. Find the voltage, current, etc.

The voltage is again 0.5 volt, the current is still 0.00125 amp, but its direction is reversed. This is true because the change which is inducing the voltage is the decreasing of the flux. Thus in order to oppose the decreasing of flux the current must reverse its direction and the polarities of A and B are likewise reversed.

12-4. Lenz's Law as an Application of the Law of Conservation of Energy. In the first example above, let us assume that the current in the wire would be counterclockwise when the flux is increasing in the downward direction. If this were the case, the induced current would in itself produce an additional increase in flux through the loop, thereby inducing more voltage and current. This leads to a contradiction of the principle of conservation of energy, for one could start the process with a small amount of external flux change; thereafter it would be self-sustaining.

12-5. Direction of the Magnetic Force on a Current. In the previous chapter it was pointed out that a force is exerted on a current or current-carrying wire owing to the interaction of the magnetic field of the current and the external magnetic field. It was mentioned, without proof, that the direction of this force could be ascertained by noting the relative directions of the lines of flux of these two fields. On one side of the current the two sets of flux lines are in the same direction and strengthen the field, while on the other side they are in opposite directions and weaken the field. The force is from the strengthened toward the weakened field.

One can arrive at this result from a consideration of Lenz's law and the law of conservation of energy. Consider the wire a-b in Fig. 12-2 to be moving from left to right across the external magnetic field directed downward into the page and represented by the xs. Let us focus our attention on a single electron e^- within this wire. As the wire moves, the electron is carried across the field from c toward d and in itself constitutes an electron current in this direction. The magnetic field about this moving electron is indicated by the tail of an arrow within a circle, $\otimes$, and the head of an arrow within a circle, $\odot$. The field is strengthened above the path of the electron and weakened below its path.

Two possibilities must be considered:

1. The force on the current is from the strengthened field toward the

weakened field. If this is true, the electron e^- will move along the wire from b toward a and through the external circuit back to b. This electron flow in the wire sets up a magnetic field about the wire indicated by $\boxed{\times}$ and $\boxed{\cdot}$. This is in accordance with Lenz's law, since the induced emf opposes the change which caused it, namely, the increase in flux through the circuit as the wire moves from left toward right.

2. Let us now examine the consequences of the opposite assumption, namely, that the force on the moving charge is from the weaker toward the stronger field. Under this assumption the electron e^- would move in

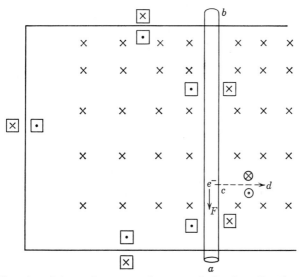

Fig. 12-2. The wire a-b is moving across the magnetic flux in a direction from left to right. The conduction electrons in the wire are forced toward end a, thus producing an electron current in a clockwise direction.

the wire from a toward b and the magnetic field associated with this electron current along the wire would be reversed from that shown in the square of Fig. 12-2. This means that the flux set up by the induced current causes an increase in total flux through the circuit and therefore aids the change which caused it. This is in contradiction to Lenz's law and the conservation of energy. It would be possible, for example, to start the wire moving from left to right with a small external force. The induced current due to this initial motion would cause a further effect in the same direction, so that one could get more energy from the system than was put into it.

12-6. Mutual Induction. Two coils (Fig. 12-3) are placed near each other but are not connected together electrically. Coil 1 is connected in series with a battery and key. The ends of coil 2 are connected to a high-impedance voltmeter. When the key in the first circuit is closed, the current rises rather rapidly and eventually reaches a value given by Ohm's law. During the build-up of this current, the flux in the surrounding region also builds up. Thus the flux through coil 2 increases with time, and a voltage is produced as indicated by the kick of the voltmeter. This process of inducing a voltage in a coil by a changing current in another coil is called mutual induction.

When the current in the first, or primary, coil has reached a steady state, the flux in the surrounding region is also constant with time, so that $d\phi_2/dt$ is now zero and the voltage induced in the secondary coil is zero (Fig. 12-4). When the key in the primary circuit is opened, the current drops, the flux decreases, and again a voltage is induced in the secondary coil but in the opposite direction (Lenz's law).

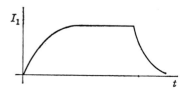

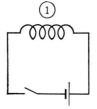

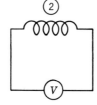

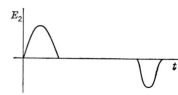

FIG. 12-3. Mutual induction. A change in current in coil 1 produces a change in flux through coil 2 and induces a voltage in this secondary coil.

FIG. 12-4. Current in the primary coil and induced voltage in the secondary coil of a mutual inductance as a function of time.

If this experiment were carried out in vacuum or at least in the absence of magnetic materials such as iron, the flux through the secondary coil would depend upon various factors such as the number of turns on the secondary coil N_2, the area of the secondary coil A_2, the value of the flux density B_1 produced by the primary coil at the secondary coil, etc. We can then write

$$\phi_2 = f(B_1, N_2, A_2, \ldots)$$

so that

$$e_2 = \frac{d\phi_2}{dt} = K\frac{dB_1}{dt}$$

Since all the factors except B_1 are independent of time, they have been lumped together into a single constant K. From Ampère's law we know that the flux density at any point in the region of a wire is strictly proportional to the current in the wire, so that $B_1 = K_1 I_1$ and

$$\frac{dB_1}{dt} = K_1\left(\frac{dI_1}{dt}\right)$$

Substituting this value yields

$$e_2 = M \frac{dI_1}{dt} \tag{12-4}$$

All the geometric constants of both circuits have been lumped together into a single constant M. M is called the coefficient of mutual induction. The name *henry* is applied to this unit, and it can be defined by Eq. (12-4). That is, if an emf of one volt is developed in the secondary circuit when the current in the primary circuit is changing at the rate of one ampere per second, the coefficient of mutual induction between the circuits is one henry.

12-7. Measurement of the Coefficient of Mutual Induction. It would appear, at first glance, that the experimental determination of the coefficient of mutual induction would be a difficult measurement, since it apparently involves the measurement of a voltage and a time rate of change of current. Most of the difficulties can be resolved, however, by using an arrangement as shown in Fig. 12-5. The secondary of the mutual inductance is connected to a ballistic galvanometer, while the primary is connected in series with a battery, a key, and an ammeter. The total resistance of the secondary circuit, including the galvanometer coil, leads, and secondary winding, is R_2 ohms. When the key in the primary is closed, the current rises from zero to I_1 amp in an unknown fashion in a time of t sec. This induces a voltage in the secondary coil equal to $e_2 = M \; (di_1/dt)$. Since the secondary

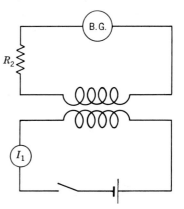

FIG. 12-5. Arrangement for measuring the coefficient of mutual inductance with a ballistic galvanometer. Conversely, this arrangement could be used to calibrate the ballistic galvanometer if the coefficient of mutual inductance were known.

circuit is closed, a current i_2 will flow during the build-up of the current in the primary. This current is

$$i_2 = \frac{e_2}{R_2} = \frac{M}{R_2} \frac{di_1}{dt}$$

The variables can be separated, and the equation can be integrated:

$$\int_0^\infty i_2 \, dt = \frac{M}{R_2} \int_0^{I_1} di_1$$

The left side is the total charge which has passed through the ballistic galvanometer Q_2 and is measured by the deflection of the galvanometer.

The right side is a constant multiplied by the change in current in the primary which is measured by the final, steady reading of the ammeter. Thus

$$M = \frac{R_2 Q_2}{I_1} \quad \text{henrys}$$

Such measurements should be made with air core and iron core inductances over a wide range of primary currents.

12-8. Self Inductance. When a current passes through a loop of wire, a magnetic field is set up everywhere in the region of the wire and magnetic lines of flux pass through the area enclosed by this wire. A change in current in the wire will therefore change the flux through the area enclosed by the wire and induce an emf in the wire itself. This is self induction. If this is carried out in the absence of magnetic material, we again have the direct proportionality between current and flux density. Following the procedure outlined in the discussion of mutual inductance, we can lump all the geometric constants into a single constant L, the coefficient of self induction (again measured in henrys), and write the induced emf as

$$e = L \frac{di}{dt}$$

The self inductance of a coil is one henry if an emf of one volt is induced in the coil when the current is changing at the rate of one ampere per second.

Self inductances of specified values can be constructed with a high degree of accuracy. The usual methods for measuring the coefficient of self inductance of a coil in the laboratory will be discussed in a later chapter on a-c bridges.

12-9. Measurement of Flux Density with a Ballistic Galvanometer. The ballistic galvanometer in conjunction with a flat coil of several turns of wire provides a simple and reliable method for measuring the flux density of a magnetic field over a wide range of values. Suppose that such a flip coil of N turns of wire encloses an effective area of A sq m and is connected to the terminals of a ballistic galvanometer. The total resistance of the circuit is R ohms. Let the coil be placed in a magnetic field with the plane of the coil parallel to the lines of flux so that none of these lines crosses the area enclosed by the coil. The coil is turned quickly through 90°, and an amount of flux equal to $B_{av}A$ webers passes through the coil. During the flipping action a voltage is induced equal to

$$e = N \frac{d\phi}{dt} = N A \frac{dB_{av}}{dt}$$

Dividing each side by R and integrating,

$$\int i \, dt = \frac{NA}{R} \int dB_{av}$$

$$Q = \frac{NA}{R} B_{av}$$

$$B_{av} = \frac{R}{NA} Q \qquad \frac{\text{webers}}{\text{sq m}}$$

The average value of the flux density (averaged over the area of the coil) is obtained in terms of the easily measured characteristics of the flip coil and ballistic galvanometer.

It is often more convenient to start with the flip coil perpendicular to the lines of flux and then turn it through 180° to produce the deflection. The flux change with respect to the coil is from $-B_{av}$ to B_{av}, so that

$$B_{av} = \frac{R}{2NA} Q \qquad \frac{\text{webers}}{\text{sq m}}$$

12-10. Fluxmeter-galvanometer. In the measurements of magnetic flux it is necessary to read the deflection of a galvanometer, and with the usual ballistic galvanometer it requires considerable expertness to read the scale just at the end of the deflection. A galvanometer having a high critical damping resistance is very greatly overdamped on a low-resistance circuit, and the free motion of the coil may be extremely slow. When magnetic flux is cutting the closed galvanometer circuit, the deflection is quick and responsive, and the galvanometer coil turns so as to cut out of the galvanometer circuit as many flux turns as are being cut into the circuit by the external flux through the test coil. When the flux ceases to change, the galvanometer coil stops moving and stands practically at rest. If the ending point of the deflection is near the natural resting point of the galvanometer, there is very little tendency for the scale reading to change after the flux has reached its steady value, and the observation is taken under stationary conditions.

12-11. Fluxmeters. Portable instruments are made that will measure the change in magnetic flux by the deflection of a pointer moving over a scale graduated in maxwells. In appearance, one of these instruments resembles a sensitive voltmeter, but in principle it is simply an overdamped ballistic galvanometer with the moving element suspended by an almost torsionless suspension so that the pointer stands at any place on the scale. When flux is measured, the pointer moves quickly from one position to another.

12-12. To Measure the Flux in an Iron Ring. From what has been said thus far it will be seen that the magnetic flux is not measured directly. It is only the *change* in the *flux turns* that affects the galvanometer and is measured by the deflection. The value of the total flux must be inferred from measurements of the change in flux. The change that is usually made in the flux through a magnetic circuit is that produced by reversing the magnetizing current. It is then assumed that the flux is reversed also and therefore that the corresponding *change* in flux is equal to twice the total value of the flux itself. This assumption is true if the residual magnetization due to larger currents has been removed from the iron by a suitable process of demagnetization.

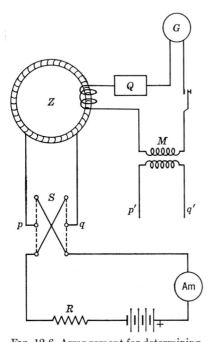

The experimental arrangement for measuring the magnetic flux in a closed-ring circuit is shown in Fig. 12-6. The iron ring is shown at Z, with the primary winding connected to the reversing switch S. If the ammeter reads the current on either side of zero, it should be placed between the ring and the reversing switch, where it will show the reversal of the current each time the switch is thrown. If the scale of the ammeter extends only on one side of the zero mark, the ammeter should be placed on the battery side of the reversing switch, in the part of the circuit where the current is in the same direction whether it is reversed through the winding on the ring or not. When S is thrown over, the current through the primary winding is reversed, thus changing the flux in the iron from ϕ in one direction to ϕ in the other direction—a change of 2ϕ. If the galvanometer remains connected to the test coil (key closed) while this change in flux occurs, there is a deflection d, and as shown before,

FIG. 12-6. Arrangement for determining the magnetic flux in the ring Z.

$$\text{Change in flux turns} = \Delta(\phi n) = 2\phi n = cd$$

where n is the number of turns in the test coil. The amount of flux in the iron before reversal is

$$\phi = \frac{cd}{2n}$$

When a ballistic galvanometer is used to measure a change in magnetic flux of $\Delta(\phi n)$ flux turns, the relation is given by

$$\Delta(\phi n) = cd$$

where the value of the constant c depends upon the resistance of the circuit in which the galvanometer is connected as well as upon the unit in which ϕ is expressed and the sensitivity of the galvanometer and its damping factor.

A known amount of magnetic flux for calibrating the galvanometer is most conveniently obtained by the use of a mutual inductance M, consisting of primary and secondary coils wound on a core of nonmagnetic material. The magnetic flux that is linked with the secondary coil, due to a current of I' amp in the primary coil, is

$$(\Phi n) = M I' \qquad \text{weber-turns}$$

when the mutual inductance M is expressed in henrys and the flux is in webers.

If the measurements are to be made in terms of maxwells ϕ instead of in webers Φ, then

$$(\phi n) = (\Phi n)10^8 = M I' 10^8 \qquad \text{maxwell-turns}$$

A reversal of the current I' will give a change in the flux turns in the secondary circuit of

$$2M I' 10^8 = cd'$$

where d' is the deflection of the galvanometer in this secondary circuit when I' is reversed in the primary coil.

The magnetic ballistic constant c for this galvanometer when connected as a part of this particular circuit is, then,

$$c = \frac{2M I' 10^8}{d'} \qquad \text{maxwell-turns/cm}$$

12-13. Relation between Magnetic Flux and MMF. In the case of a magnetic circuit of iron or other ferromagnetic material, the magnetic flux is not linearly proportional to the mmf. This is

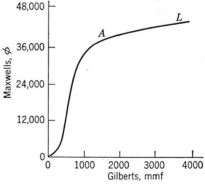

Fig. 12-7. Magnetization curve for a circuit of tool steel 50 cm in length and 3 sq cm in cross section.

shown by measuring the flux in the iron ring that corresponds to different values of the magnetizing current and then plotting the results as a curve (Fig. 12-7).

It is seen that for small values the flux increases faster than the mmf, while for larger values the flux increases more and more slowly, although the curve never reaches a maximum however much the mmf is increased.

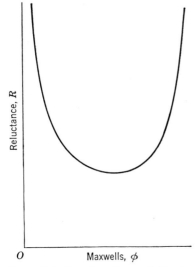

FIG. 12-8. Showing the variation in the reluctance of a magnetic circuit as the flux is increased.

The effect of previous magnetization should be removed before these measurements of the magnetic flux are made.

12-14. The Reluctance of a Magnetic Circuit. Another way of showing the relations in a magnetic circuit is to plot the values of the reluctance of the circuit corresponding to different values of the flux. This does not require any new measurements, since the values of the reluctance R are computed from the relation

$$R = \frac{\text{mmf}}{\phi}$$

and the same values of ϕ and mmf are used that were obtained before for the curve in Fig. 12-7. The reluctance is less for medium values of the flux than it is for small or large values. The minimum reluctance corresponds to a value of the flux of about half saturation.

12-15. Demagnetization. *Hysteresis.* As stated above, the reversal of the magnetizing current does not ensure a change in the value of the flux from ϕ in one direction to an equal amount in the opposite direction (giving a change of 2ϕ) unless the effect of any stronger magnetization has been removed from the iron. To obtain this essential demagnetization requires an understanding of the condition of the iron and the careful manipulation of the rather elaborate process of demagnetization.

When a ring of steel is magnetized by a current that starts at zero and steadily increases to a final value, the resulting flux density increases as shown by the curve in Fig. 12-7. But now suppose that when the point A on the curve has been reached, the current is *decreased* gradually to zero. Would the magnetization curve be retraced, or

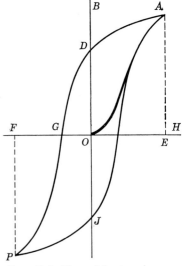

FIG. 12-9. Normal hysteresis curve for steel.

would a new curve be obtained? Investigation reveals that as the current is slowly decreased, the magnetization of the steel does not decrease to its former values, and when the current reaches zero a large amount of "residual" magnetization still remains. This decreasing magnetization is shown by the curve AD (Fig. 12-9). The ordinate OD shows the residual flux density when the current has been decreased to zero.

It will even require the application of a reversed magnetic intensity, equal to OG, to reduce the magnetic flux density to zero. This negative value of H is called the "coercive force" of the steel. It has a large value for hard iron and steel but is small for soft iron and silicon-iron alloys. If the reversed H is increased to a value OF, equal and opposite to OE, the material is found to be magnetized as strongly as before but in the opposite direction, and it will hold this magnetization just as persistently as the other. If H is reduced to zero and again increased to OE, the flux density follows as shown by the curve PJA. This lagging of the values of B behind the corresponding changes in H is called "hysteresis," from a Greek word meaning "to lag behind." The complete curve as thus drawn between B and H is called a "hysteresis curve."

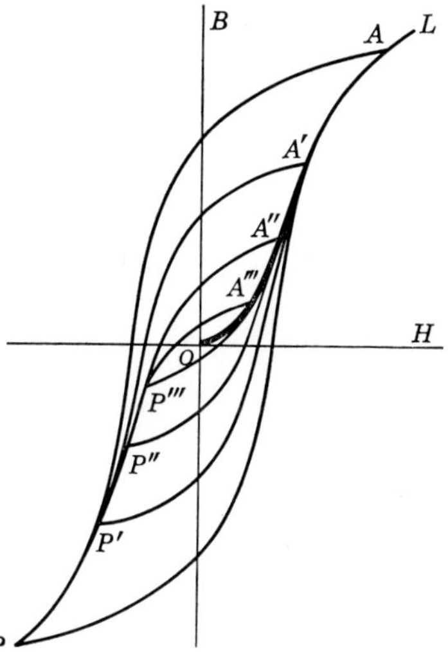

There is not just one hysteresis curve for a given kind of iron. The curve shown in Fig. 12-9 was traced by reducing the value of H from the point A. A similar curve can be drawn starting from any other point on the curve OAL

FIG. 12-10. Showing a few of the hysteresis loops through which a bar of steel is carried when it is being demagnetized.

(Fig. 12-7), and a series of such curves are shown in Fig. 12-10. The upper tips of all these hysteresis loops rest upon the curve OAL; in fact, the standard method for finding the magnetization curve OAL is to determine the location of the tips of various hysteresis curves.

12-16. Unsymmetrical Hysteresis Curves. In the case shown by Fig. 12-9 the magnetic intensity H was varied between equal positive and negative values, and the corresponding changes in the flux density resulted in the symmetrical curve shown there. Other ways of changing H would result in other forms of hysteresis curves. If the steel has been magnetized to the extent shown by the point A and then the current is gradually reduced to zero, the steel will be left at D. In a closed-ring circuit the steel will hold this magnetization for a long time.

If after H has been reduced to zero it is increased again in the same direction, the steel will be carried back to A along a path like DSA (Fig. 12-11). By making and breaking the magnetizing current, the steel can be taken around the loop $DSAD$ as many times as desired, but the magnetization is never reversed or reduced to a value smaller than OD.

If when the steel is at D a small value of H is applied and varied between equal positive and negative values, the steel will be carried

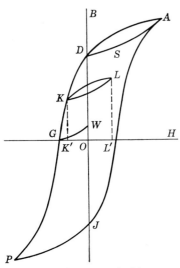

around a small loop like KL. No matter how many times this small value of H is reversed, the values of B are always in the same direction and merely vary between the values of KK' and LL', and after many reversals the steel is magnetized nearly as strongly as before.

If the magnetic flux is reduced to zero by bringing the steel to a point like G, the steel is not demagnetized, for when H is reduced to zero, B increases to OW. By varying H between zero and OG, the steel can be carried around the small loop GW as often as desired. Even if the steel should be carried to a point slightly below G, so that when H is reduced to zero the curve would come back to O, the steel would

FIG. 12-11. Unsymmetrical hysteresis loops.

not be in the unmagnetized condition but would respond differently to a small negative value of H than to an equal positive value.

There is no limit to the number of curves that might be traced, and when a piece of steel or iron has been subjected to various unknown changes in H, there is no telling in what condition it may be left.

12-17. Demagnetization of Iron or Steel. When a ring of iron or steel has been magnetized by a current, the material retains a large part of the magnetization after the current is stopped. This residual magnetization must be removed before the normal magnetic qualities of the material can be studied. As shown above, it is not easy to leave the iron unmagnetized when the current is removed. Simply stopping the current has little effect. Reversing the current leaves the iron as strongly magnetized in the other direction. But by carefully following the process of demagnetization that is outlined below it is possible to reduce the magnetization gradually and to leave the iron as nearly demagnetized as may be required.

When a current that is as large as any that has been used previously is reversed a few times in the primary winding of the ring, the iron is carried around a symmetrical hysteresis curve. If this current is reduced slightly and reversed a few more times, the first reversal will not carry the iron to an equal negative value of B, but after several reversals the iron will be carried around a symmetrical loop that is slightly smaller than the former loop. If the decrease in current has been too great, the resulting hysteresis loop will be unsymmetrical, as shown above.

Many reversals are desirable, but having regard for the labor and time involved, 10 reversals are usually sufficient, especially if they are slow enough to allow the eddy currents to die away after each reversal. This method requires smaller steps and slower reversals on the steep part of the B-H normal magnetization curve.

After the iron has been reduced from the first hysteresis curve to a symmetrical loop that is slightly smaller, the same process can be used again to bring the iron to another cycle that is still smaller. By repeating this process again and again, the magnetization can be gradually worked down to as small a value as may be required. Usually it is safe to reduce the current by steps of about 10 per cent of the maximum value, until the point of maximum permeability of the iron has been reached, and then to use steps that will decrease the flux density by about 1,000 gauss at each step.

When reducing the current it is somewhat better to make the reductions alternately at opposite ends of the hysteresis loop, $i.e.$, first reduce the positive value of H, and next reduce H when it is negative, and the next time when it is positive, and so on. This helps to keep the decreasing loops symmetrical with respect to the axes of B and H.

12-18. Normal Magnetization Curve. If it were possible to obtain a ring or bar of iron that never had been near a current or any magnetized steel, the relation between B and H could be studied by gradually increasing H from zero to larger values and measuring the corresponding values of B. But after the iron has once been magnetized, it is necessary to demagnetize it before making measurements on the relation between B and H. The standard curve for showing the magnetization of iron is obtained after the material is thoroughly demagnetized. In order to locate a point on this curve, the current that magnetizes the iron is reversed many times until the iron is in a symmetrical cyclic condition corresponding to a hysteresis curve like one of those shown in Fig. 12-10. Then the galvanometer is read at the time of the next reversal of the current. The deflection measures the total height of the corresponding hysteresis loop, and half of this deflection measures the height of the tip of the loop above the H axis. A series of such points would locate the tips of the loops shown in Fig. 12-10, and the curve through these points is called the "normal magnetization curve" for the given substance. This curve is nearly like the one that would have been obtained by gradually increasing the magnetization the first time that the iron was ever magnetized.

12-19. Normal Magnetization Determination during Demagnetization.
It makes little difference in the normal magnetization curve whether the iron is first demagnetized down to nearly zero and the values of B are then determined for increasing values of H or the points on the curve are determined in the opposite order. At any stage in the process of demagnetization, after many reversals with a given current I, the iron is magnetized as strongly as though the demagnetization had been completed and the current I then applied and reversed a few times. If the deflection of the galvanometer is read for the last reversal of current before the next smaller value is considered, it will measure the normal value of B corresponding to that value of the current. It is therefore possible to take the readings for a curve of normal magnetization during the process of demagnetization, and with the saving of considerable time and effort.

12-20. Permeameters. The standard method for magnetic testing is the ring method, since a ring circuit is uniform throughout its length and there are no ends to introduce disturbing factors. Because of the difficulty of making rings and the labor of carefully winding each one, many other arrangements have been used in which the coils are wound on permanent forms that can be slipped over a bar or rod of iron. The various forms of permeameters have magnetic circuits that more or less satisfactorily approach the form of a ring circuit.

12-21. The Double-bar Permeameter. In the double-bar permeameter shown in Fig. 12-12, the magnetic circuit consists of two rectangular

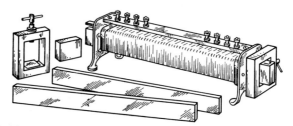

Fig. 12-12. Smith permeameter for obtaining the normal magnetization and hysteresis curves for iron or steel.

bars, 1 by 3 cm in cross section and 36 cm in length. At each end these bars are connected by a flat block of Armco iron. The sides of the bars and the soft iron blocks are finished to a smooth surface to ensure good contact between them. The bars are held firmly against the iron blocks by a yoke provided with a screw clamp that does not mar the smooth surface of the bars when they are clamped together. The magnetic circuit is thus closed, as shown in Fig. 12-13, and consists almost entirely of the bars that are being tested.

The bars thread through two brass tubes of rectangular section, slightly larger than the bars, mounted on end brackets and upon which are wound three layers of well-insulated wire. Considerable care is exercised to ensure the uniform spacing of the turns of wire in these coils, and the number of turns per centimeter is constant along the length of the greater part of each bar. A part of the magnetic circuit lies beyond the ends of the coils, where it would be practically impossible to wind the coils. The reluctance of this end portion of the magnetic circuit includes the reluctance of the iron block and the contact reluctance between the block and bars. These reluctances are made small by having the iron block short and wide. In addition there is the reluctance of a short length of the bars and an equivalent further length of the same material for the path of the flux as it turns to reach the side of the bar. This extra length is kept small by making the bars broad and thin. The total reluctance of the end portion is thus largely due to the material of the bars and is therefore a definite fraction of the reluctance of the entire circuit. This fraction is determined experimentally when the permeameter is designed, and the number of turns that are required for the end portions of the magnetic

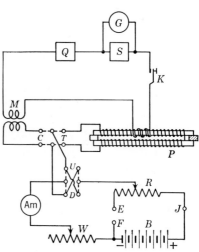

Fig. 12-13. Normal magnetization of the iron bars in the permeameter P.

circuit are wound as extra turns at each end of both long coils. The two long solenoids and the four extra end coils are all joined in series, forming a single electric circuit with the ends brought out to a pair of binding posts. The magnetic flux is measured by means of a small test coil of 20 turns, wound on the brass tubes before the current winding is put on. Ten turns are distributed over the middle third of each bar, thus giving an average of flux density measurement for both bars that is not affected very much by small irregularities at the ends of the bars.

After one pair of bars is tested, the bars of another grade of iron or steel can be inserted easily and tested without disturbing the test coils or the solenoids that carry the magnetizing current.

12-22. Circuit Connections. *Using the Double-bar Permeameter.* The experimental arrangement is shown in Fig. 12-13. The permeameter is shown at P with its primary winding connected to the double throw switch CT. When this switch is closed on T, connection is made to the reversing switch UD. The latter switch should be of the quick-reversing type (Fig. 12-14) that will change the current through P from its full value in one direction to the same value in the opposite direction before the galvanometer has moved more than a small part of its full deflection. By tracing the circuit through this switch it is seen that only one of the cross connections is in the circuit at one time, and this keeps the same resistance in the circuit whether the switch is closed at U or at D. A battery of 20 to 40 volts is desirable to swamp the emf induced in the

Fig. 12-14. A simple switch for reversing a current quickly. (*Courtesy of Leeds & Northrup Co.*)

primary of P and to compel a quick rise of the current after reversal. Two rheostats, W and R, are shown. These should be able to carry the maximum current that will be used, perhaps 2 or 3 amp, and should have perhaps 50 ohms each. When small currents are needed, a very large resistance is required for R, unless a small part of the battery can be used. By connecting E and F, the rheostat EJ becomes a potential divider across the battery, and the current through P can be made as small as required by moving the sliding contact toward E, with W kept large. Care should be taken to make sure that the resistance of EJ is large enough to be safely connected across the battery and that the current through the part R does not exceed the safe carrying capacity of the rheostat.

The potential divider is not desirable with large currents in P, as it puts an extra heating load on the rheostat EJ.

12-23. The Hysteresis Curve. The hysteresis curve for steel is shown by the loop $UNDJU$ in Fig. 12-15. With each reversal of the magnetizing current the steel is carried around the hysteresis cycle, but in order to trace the form of this curve it is necessary to be able to stop at various points along the curve by changing the magnetic intensity H in steps that are smaller than a complete reversal.

12-24. Standard Point of Reference. Since only changes in B can be measured and not the full value of B itself, it is necessary to choose one point on the hysteresis curve as a point of reference and then measure the change in B when the steel is carried from any condition to this reference point. When the steel is in the form of thin sheets or fine wires, it makes little difference whether point $N(H = 0)$ or point D (Fig. 12-15) is chosen for this reference point. But for solid bars of steel there are eddy currents in the body of the metal whenever B is changed, and the magnetizing effect of these eddy currents is added to the value of H. Thus B does not reach its final value until the eddy currents die away. Because of this effect it has been shown that the change in B can be measured more accurately when the change ends with a large value of H. There-

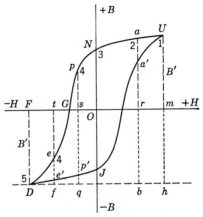

Fig. 12-15. Hysteresis curve for a sample of ferromagnetic material. $H = 0.4\pi nI$ oersteds or nI amp-turns/m. $B = \phi A$ gauss or webers/sq m.

fore it is better to choose the tips D and U of the hysteresis curve as the points of reference for locating the other points of the curve.

12-25. Location of Points on the Hysteresis Curve. The accurate tracing of a hysteresis curve by the ballistic galvanometer method involves many careful manipulations in order to make certain that the sample is at all times maintained in its cyclic state. It is possible to carry out the various steps by the use of several switches and keys and obtain good results. Several specialized switches have been devised to help avoid errors in the manipulations.

One such arrangement (Fig. 12-16) makes use of a modified double-pole double-throw switch. The modification consists of elongating two of the fixed contacts indicated by Y and Z. Both sides of the knife-blade arrangement are connected together and tied to one end of the primary coil of the permeameter. The center of the battery is connected to the other end of the primary coil.

When the switch is thrown fully to the right, the sample is at point U

(Fig. 12-15). The switch is then raised so that contact at X is broken but contact at Y is still maintained. This suddenly throws the resistance $R+$ in series with the circuit and changes the sample to position a on the hysteresis curve. The deflection of the galvanometer yields the change in flux which has occurred.

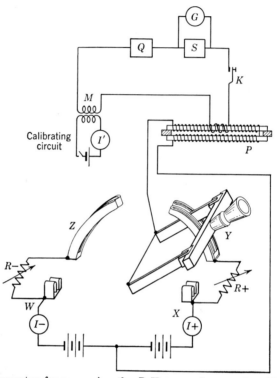

Fig. 12-16. Apparatus for measuring the B-H curve of a sample. The switch is a simple double-pole double-throw type which has been modified. The secondary of the calibrating mutual inductance M remains in the galvanometer circuit during the measurements.

The switch is then further advanced to make contact at Z alone. This puts the sample at position e in Fig. 12-15. When the switch is further depressed to make contact at terminal W, the resistance $R-$ is shorted out and the sample is carried to the lower tip D of the hysteresis curve. Again the reading of the deflection of the galvanometer yields the change in flux between e and D.

The process is continued; in raising the knife blade to break contact at W while maintaining contact with the elongated terminal Z, the sample is carried to the point e' on the lower half of the curve, etc. If the resistances $R+$ and $R-$ are varied, additional points on the B-H curve can be obtained.

PROBLEMS

12-1. A horizontal wire lying in the north-south direction is moved toward the east with a velocity of 20 m/sec. The wire is 3 m long. Find the voltage induced in the wire, in magnitude and direction, if the magnetic field in this region is directed downward and the flux density is 0.1 weber/sq m.

12-2. The ends of the wire in Prob. 12-1 are connected to a 2-ohm resistance external to the magnetic field. Find the force required to keep the wire moving at a constant speed of 20 m/sec.

12-3. Express the henry in terms of mass, length, and time.

12-4. A solenoid 2 m long and 0.1 m in diameter is wound uniformly with 1,000 turns of wire. Calculate the coefficient of self induction of this coil.

12-5. A second coil of five turns of wire is wound at the center of the solenoid of Prob. 12-4. Calculate the coefficient of mutual induction between these two coils.

12-6. A coil with an area of 20 sq cm having five turns of wire is placed in a magnetic field with the plane of the coil at right angles to the direction of the lines of flux. The coil is connected to a ballistic galvanometer, and the total circuit resistance is 50 ohms. Find the charge flowing through the galvanometer when the coil is removed from the magnetic field if the average flux density is 12,000 gauss.

12-7. The primary coil of a mutual inductance has a self inductance of L_1 henrys. The secondary coil has a self inductance of L_2 henrys. The mutual inductance between coils is M henrys. The primary winding is connected in series with the secondary to form a single self inductance of L henrys. Prove that $L = L_1 + L_2 + 2M$ when the coils are connected to aid each other and $L = L_1 + L_2 - 2M$ when connected in opposition.

12-8. Explain the hysteresis curve in Fig. 12-15 in terms of the alignment of the permanent magnetic dipoles of the atoms.

CHAPTER 13

TRANSIENT CURRENTS

13-1. Introduction. The majority of the circuits which have been discussed in earlier chapters have been made up of simple elements such as batteries and resistors. This led to conditions yielding currents which are independent of time. While treating the ballistic galvanometer in Chap. 6, it was found that a capacitor charged from a battery in series with a resistance produced time-dependent or transient currents. An investigation of such transient effects gives an introduction to the transition region between d-c and a-c techniques.

13-2. Growth of a Current through an Inductance. Figure 13-1 represents a series circuit consisting of a battery, key, resistance, and inductance. When the key is closed, the current will rise at a rate of di/dt amp/sec. We have seen previously that such a current change through an inductance L will, in itself, produce an additional source of voltage of magnitude $L\,di/dt$ in such a direction as to oppose the changing of the current. Kirchhoff's second law can be applied to this circuit at any instant of time:

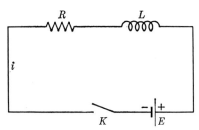

Fig. 13-1. Simple inductive circuit. The resistance R represents the entire resistance of the circuit.

$$L\frac{di}{dt} + iR = E$$

When the variables are separated, the equation can be integrated immediately:

$$\frac{di}{E - iR} = \frac{1}{L}\,dt$$

$$\ln\,(E - iR) = -\frac{R}{L}t + K$$

The constant of integration K can be evaluated if we recall that at the instant the key is closed, the current in the circuit is zero. Thus

204

$K = \ln E$, and the equation can be written

$$\ln \frac{E - iR}{E} = -\frac{R}{L} t$$

Since the logarithm of a number is defined as the power to which the logarithmic base must be raised to equal the number, we can write

$$\frac{E - iR}{E} = \epsilon^{-(R/L)t}$$

$$i = \frac{E}{R} [1 - \epsilon^{-(R/L)t}] \qquad (13\text{-}1)$$

The equation is plotted in Fig. 13-2. It is seen that the current rises at a finite rate and after a long period of time approaches the value $I = E/R$.

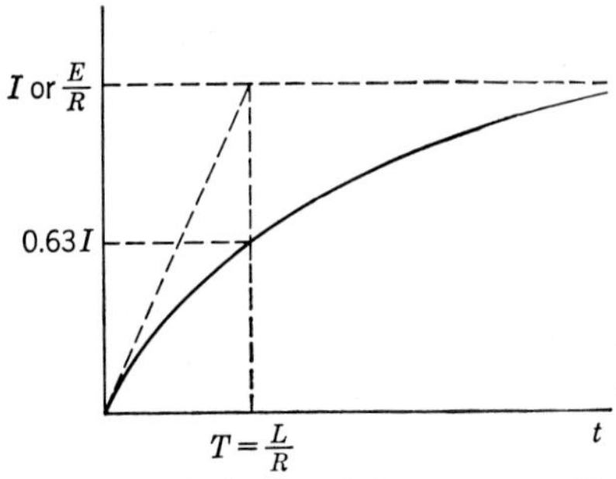

FIG. 13-2. Variation of the current with time for an inductive circuit. The time constant is T.

13-3. Time Constant for an LR Circuit. From the results obtained above, one may be concerned about measurements taken with d-c meters, since we have shown that current may change with time even though the source of current is a constant-voltage battery. This question should be considered because even a very simple circuit will contain a certain amount of inductance along with some resistance. The problem can be stated: After closing a key to complete a circuit, how long must one wait before taking a meter reading? To give the order of magnitude of this time we define a circuit constant called the time constant. This is the time which would be required for the circuit to reach its final state if the measured quantity were to continue to change at its initial rate. For the circuit in Fig. 13-1 this can be written $T = I/(di/dt)_{t=0}$ sec. This is illustrated on the curve of Fig. 13-2. The quantity $di/dt = (E/L)\epsilon^{-(R/L)t}$, and its value at $t = 0$ is $(di/dt)_{t=0} = E/L$. Thus for the LR circuit the time constant is L/R sec when L is expressed in henrys and R in ohms.

The time constant provides a simple method for estimating the order of magnitude of the rise time of the current. For example, consider a circuit made up of a 10-ohm resistance in series with a 10^{-3}-henry inductance. The time constant is 10^{-4} sec, and for all practical purposes, the current can be considered to have reached its final value in a few time constants, say 10^{-3} sec. On the other hand, a magnet coil may have a resistance of 0.5 ohm and an inductance of 20 henrys. The time constant is 40 sec, and it would be necessary to wait for several minutes before a meter reading would give the ultimate value of the current.

The actual value of the current after one time constant can be found by substituting the value L/R in Eq. (13-1). $i/I = (1 - 1/\epsilon) \approx 0.63$; that is, the actual current has risen to 63 per cent of its ultimate value after one time constant.

13-4. Energy Distribution in an Inductive Circuit. During the build-up of the current through an inductive circuit, the battery is supplying energy at the rate of Ei joules/sec. Energy is being converted into heat in the resistance at the rate of i^2R joules/sec. These two quantities are not equal. Their difference represents the rate at which energy is being stored in the magnetic field of the inductance L. It is of interest to evaluate this stored energy in terms of the various circuit quantities.

The total energy supplied by the battery is

$$W_{\text{battery}} = \int_0^t Ei\, dt = \int_0^{t \to \infty} EI[1 - \epsilon^{-(R/L)t}]\, dt$$
$$= EIt - \frac{EIL}{R}$$

The energy converted into heat in the resistance is

$$W_{res} = \int_0^t i^2R\, dt = \int_0^{t \to \infty} I^2R[1 - 2\epsilon^{-(R/L)t} + \epsilon^{-(2R/L)t}]\, dt$$
$$= EIt - \frac{3}{2}\frac{EIL}{R}$$

The energy stored in the inductance is found to be

$$W_{ind} = W_{\text{battery}} - W_{res} = \tfrac{1}{2}LI^2 \qquad \text{joules} \qquad (13\text{-}2)$$

This energy is not dissipated, but rather, it is stored in the magnetic field of the inductance. When the current is decreased, the magnetic field decreases and the stored energy is again made available as electrical energy.

13-5. Decay of Current in an Inductive Circuit. After the key in Fig. 13-1 has been closed for many time constants, the current will have approached its limit of I amp. Let us now imagine that the battery can be removed effectively from the circuit without breaking the circuit. (This might be accomplished by leaving the battery in the circuit but connecting its terminals with a wire of zero resistance.) Kirchhoff's law can be written

$$L\frac{di}{dt} + iR = 0$$

The variables can again be separated, and the integration can be carried out.

$$\ln i = \frac{-R}{L}t + k'$$

If we start counting time when the effect of the battery is removed, the initial conditions can be written: at $t = 0$, $i = I$. Thus $k' = \ln I$ and

$$i = I\epsilon^{-(R/L)t} \qquad (13\text{-}3)$$

The current decays exponentially with time as indicated in Fig. 13-3. The voltage and energy are supplied by the inductance during this decay process.

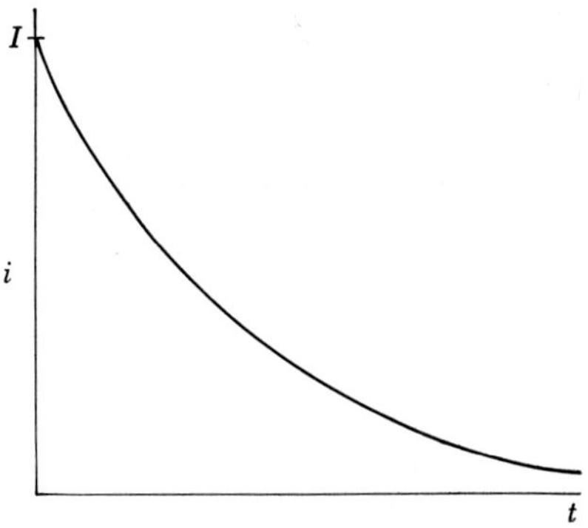

13-6. Charging and Discharging of a Condenser. Solutions have been given in Chap. 6 for the charge on a condenser as a function of time. The circuits for charging and discharging are indicated in Fig. 13-4. During the charging process we have found that

$$q = Q[1 - \epsilon^{-(1/RC)t}] \qquad (13\text{-}4)$$

The discharge takes place exponentially with time and can be written

$$q = Q\epsilon^{-(1/RC)t} \qquad (13\text{-}5)$$

Since current is the rate of flow of charge, the transient currents during charging and discharging can be obtained by differentiating Eqs. (13-4) and (13-5). During the charging process

$$i = \frac{dq}{dt} = \frac{Q}{RC} \epsilon^{-(1/RC)t} = I\epsilon^{-(1/RC)t} \qquad (13\text{-}6)$$

During the discharge

$$i = -I\epsilon^{-(1/RC)t} \qquad (13\text{-}7)$$

The values of q and i are plotted in Fig. 13-5.

Fig. 13-3. Decay of current in an inductive circuit.

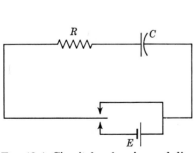

Fig. 13-4. Circuit for charging and discharging a condenser through a resistance.

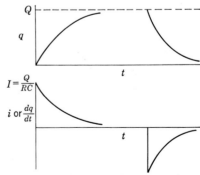

Fig. 13-5. Charge and current in a capacitive circuit during the charging and discharging process.

The time constant for the circuit is

$$T = \frac{Q}{(dq/dt)_{t=0}} = RC \qquad \text{sec} \qquad (13\text{-}8)$$

It may be noted that the time constant is a characteristic of the circuit and remains the same when the build-up of charge or current is considered as well as during the decay of these quantities.

13-7. Energy Stored in a Charged Condenser. The discharge current for an RC circuit is given by Eq. (13-7). The power expended in the resistance during discharge is i^2R. The total amount of energy delivered to the resistance by the condenser is

$$W_{cond} = \int_0^\infty i^2R \, dt = \tfrac{1}{2}I^2R^2C = \tfrac{1}{2}CV^2 \qquad \text{joules} \qquad (13\text{-}9)$$

13-8. Discharge of a Condenser through an Inductance and Resistance. After the condenser in Fig. 13-6 has been given a charge Q, the key is closed to complete the circuit. The time variation of the charge on the condenser and the current in the circuit can be obtained by applying Kirchhoff's second law:

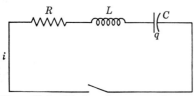

$$L \frac{di}{dt} + Ri + \frac{q}{C} = 0$$

Since $i = dq/dt$,

FIG. 13-6. Discharging a condenser through an inductance and a resistance in series.

$$\frac{di}{dt} = \frac{d^2q}{dt^2}$$

Then the equation in terms of q becomes

$$\frac{d^2q}{dt^2} + \frac{R}{L}\frac{dq}{dt} + \frac{q}{LC} = 0 \qquad (13\text{-}10)$$

This second-order differential equation is discussed thoroughly in any text on differential equations. The solution can be written

$$q = A \exp\left[\left(-\frac{R}{2L} + \sqrt{\frac{R^2}{4L^2} - \frac{1}{LC}}\right)t\right]$$

$$+ B \exp\left[\left(-\frac{R}{2L} - \sqrt{\frac{R^2}{4L^2} - \frac{1}{LC}}\right)t\right]$$

The constants of integration A and B can be evaluated from the initial conditions which exist in the circuit. These conditions can be written as follows: At $t = 0$, the charge $q = Q$; also at $t = 0$, the current

$$\left(\frac{dq}{dt}\right)_{t=0} = 0$$

The latter condition is evident from the solution of the circuit consisting of an inductance, resistance, and battery (Sec. 13-2). At the instant that the key is closed, the capacitor plays the role of a battery of voltage Q/C.

There are three separate phases of the general solution which must be considered in detail and which depend upon the relative values of R, L, and C. The first case to be considered is that in which $R^2/4L^2 > 1/LC$. The initial conditions in the circuit can be applied in order to evaluate A and B.

At $t = 0$: $\qquad\qquad q = Q \qquad$ or $\qquad A + B = Q$

The second condition yields

$$\left(\frac{dq}{dt}\right)_{t=0} = A\left(-\frac{R}{2L} + \sqrt{\frac{R^2}{4L^2} - \frac{1}{LC}}\right) + B\left(-\frac{R}{2L} - \sqrt{\frac{R^2}{4L^2} - \frac{1}{LC}}\right) = 0$$

A solution can now be obtained for both A and B. For convenience of notation let us call $R/2L = \alpha$ and $\sqrt{(R^2/4L^2) - (1/LC)} = \beta$. Then

$$A = \frac{Q(\alpha + \beta)}{2\beta} \qquad \text{and} \qquad B = \frac{Q(-\alpha + \beta)}{2\beta}$$

The complete equation for the charge on the condenser as a function of time is

$$q = \frac{Q(\alpha + \beta)}{2\beta} \epsilon^{(-\alpha+\beta)t} + \frac{Q(-\alpha + \beta)}{2\beta} \epsilon^{(-\alpha-\beta)t} \qquad (13\text{-}11)$$

The charge gradually leaks off the condenser as shown in Fig. 13-7. The role of the individual exponential terms is also indicated.

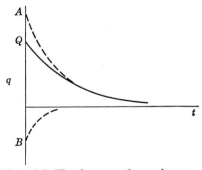

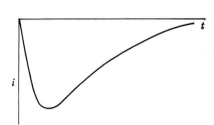

FIG. 13-7. The charge on the condenser as a function of time. The circuit is over-damped. The relative influences of the two exponential terms is indicated.

FIG. 13-8. Current in the overdamped circuit.

The variation of current with time is also evident (Fig. 13-8):

$$i = \frac{dq}{dt} = \frac{Q(\alpha + \beta)(-\alpha + \beta)}{2\beta} \epsilon^{(-\alpha+\beta)t} + \frac{Q(-\alpha + \beta)(-\alpha - \beta)}{2\beta} \epsilon^{(-\alpha-\beta)t}$$

$$= \frac{V}{2L\beta}[\epsilon^{(-\alpha-\beta)t} - \epsilon^{(-\alpha+\beta)t}] \qquad (13\text{-}12)$$

V is the initial potential of the condenser Q/C.

When $R^2/4L^2$ approaches $1/LC$, the quantity β approaches zero. Since β occurs in the denominator of our solution, care must be taken in evaluating this equation as $\beta \to 0$. The values of the charge and current for this special case, $R^2/4L^2 = 1/LC$, can be found by expanding Eqs. (13-11) and (13-12) and allowing β to approach zero:

$$
\begin{aligned}
q &= \frac{Q\alpha}{2\beta}\,\epsilon^{-\alpha t}\epsilon^{\beta t} + \frac{Q}{2}\,\epsilon^{-\alpha t}\epsilon^{\beta t} - \frac{Q\alpha}{2\beta}\,\epsilon^{-\alpha t}\epsilon^{-\beta t} + \frac{Q}{2}\,\epsilon^{-\alpha t}\epsilon^{-\beta t} \\
&= \frac{Q}{2}\,\epsilon^{-\alpha t}(\epsilon^{\beta t} + \epsilon^{-\beta t}) + \frac{Q\alpha}{2}\,\epsilon^{-\alpha t}\left(\frac{\epsilon^{\beta t} - \epsilon^{-\beta t}}{\beta}\right) \\
&= \frac{Q}{2}\,\epsilon^{-\alpha t}\left[1 + \beta t + \frac{\beta^2 t^2}{2!} + \cdots + \left(1 - \beta t + \frac{\beta^2 t^2}{2!} + \cdots\right)\right] \\
&\quad + \frac{Q\alpha}{2\beta}\,\epsilon^{-\alpha t}\left[1 + \beta t + \frac{\beta^2 t^2}{2!} + \cdots - \left(1 - \beta t + \frac{\beta^2 t^2}{2!} - \cdots\right)\right] \\
q &= Q\epsilon^{-\alpha t} + Q\alpha t\epsilon^{-\alpha t} \qquad \lim \beta \to 0 \\
&= Q\left(1 + \frac{R}{2L}\,t\right)\epsilon^{-(R/2L)t} \tag{13-13}
\end{aligned}
$$

With $\beta = 0$, the circuit is said to be critically damped, and it will approach its equilibrium state in a minimum time.

The current for the critically damped case is obtained by differentiating Eq. (13-13).

$$
\begin{aligned}
i = \frac{dq}{dt} &= -Q\alpha\epsilon^{-\alpha t} - Q\alpha^2 t\epsilon^{-\alpha t} + Q\alpha\epsilon^{-\alpha t} \\
&= -Q\alpha^2 t\epsilon^{-\alpha t}
\end{aligned}
$$

Since $\alpha^2 = R^2/4L^2$ and in this case $R^2/4L^2 = 1/LC$,

$$
i = -\frac{V}{L}\,t\epsilon^{-(R/2L)t}
$$

Again V is the initial voltage on the condenser.

The third condition to be considered arises when $R^2/4L^2 < 1/LC$. The familiar term $\beta = \sqrt{(R^2/4L^2) - (1/LC)}$ then becomes imaginary, and while this does not interfere with the correctness of the previous solution, it does not lend itself to an easy physical interpretation in the exponential form. It is possible, however, to express the complex exponentials in the form of sine and cosine functions and obtain the results

$$
q = D\epsilon^{-\alpha t}\cos(\gamma t + \phi)
$$

The quantity $\gamma = \sqrt{(1/LC) - (R^2/4L^2)}$ is a real number, while D and ϕ are the two constants which can be determined from the initial conditions.

Applying the initial conditions

At $t = 0$, $q = Q$: $\quad D = \dfrac{Q}{\cos \phi}$

At $t = 0$, $i = 0$: $\quad 0 = -D\gamma \sin \phi - D\alpha \cos \phi$

Thus ϕ is an angle whose tangent is $-\alpha/\gamma$ and $\cos \phi = \sqrt{1 - (CR^2/4L)}$. The full equation for charge as a function of time is

$$q = \frac{Q\epsilon^{-\alpha t}}{\sqrt{1 - (CR^2/4L)}} \cos\left(\gamma t - \tan^{-1} \frac{\alpha}{\gamma}\right) \qquad (13\text{-}14)$$

This equation is plotted in Fig. 13-9. It may be noted that the charge oscillates back and forth through the circuit with an amplitude which decreases exponentially with time. In the limit, as R approaches zero, the curve takes the form of a pure, undamped cosine wave. The energy in the system is continually interchanged between the condenser and inductance.

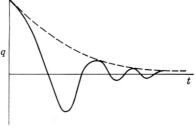

FIG. 13-9. Charge on a condenser as a function of time. The circuit is underdamped and oscillates with a period $T = \dfrac{2\pi}{\sqrt{(1/LC) - (R^2/4L^2)}}$.

13-9. Charging a Capacitor through an Inductance and Resistance. A capacitor is to be charged from a battery of voltage E using the circuit indicated in Fig. 13-10. Kirchhoff's second law yields the relationship between the various quantities as follows:

$$L \frac{di}{dt} + Ri + \frac{q}{C} = E$$

or

$$\frac{d^2q}{dt^2} + \frac{R}{L} \frac{dq}{dt} + \frac{q}{LC} = \frac{E}{L} \qquad (13\text{-}15)$$

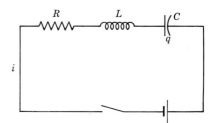

FIG. 13-10. Charging a condenser through an inductance and resistance in series.

Equation (13-15) can be reduced to the form of Eq. (13-10) by the usual technique of changing variables. Let $q = y + CE$. (CE is actually the maximum charge Q which the condenser could obtain from the battery alone.)

$$\frac{d^2y}{dt^2} + \frac{R}{L} \frac{dy}{dt} + \frac{y}{LC} = 0$$

For large damping $R^2/4L^2 > 1/LC$ (and again setting $R/2L = \alpha$ and $\sqrt{(R^2/4L^2) - (1/LC)} = \beta$)

$$y = A_1\epsilon^{(-\alpha+\beta)t} + B_1\epsilon^{(-\alpha-\beta)t}$$

Thus $$q = Q + A_1\epsilon^{(-\alpha+\beta)t} + B_1\epsilon^{(-\alpha-\beta)t}$$

The initial conditions determining the constants of integration are at $t = 0$, $q = 0$ and at $t = 0$, $i = 0$. (The latter condition can be understood if one realizes that when the key is first closed, the condenser has zero charge and zero voltage. At this instant it can have no effect on the circuit, which, therefore, will act exactly like a battery in series with only a resistance and inductance.) The complete equation for the charging process is

$$q = Q - \frac{Q}{2}\frac{\alpha+\beta}{\beta}\epsilon^{(-\alpha+\beta)t} - \frac{Q}{2}\frac{-\alpha+\beta}{\beta}\epsilon^{(-\alpha-\beta)t} \qquad (13\text{-}16)$$

The equation indicates that the charge simply leaks onto the condenser in the manner shown in Fig. 13-11 and never exceeds the value CE.

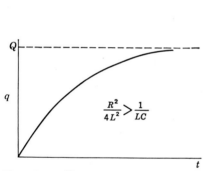

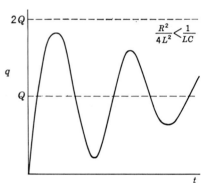

FIG. 13-11. Charge on a condenser as a function of time for an overdamped circuit.

FIG. 13-12. Charge on a condenser as a function of time for an underdamped circuit.

When the values of the circuit components are such that critical damping exists ($R^2/4L^2 = 1/LC$), the exponentials can again be expanded and evaluated as in Sec. 13-8 and yield

$$q = Q - Q(1 + \alpha t)\epsilon^{-\alpha t}$$

When the circuit is underdamped ($R^2/4L^2 < 1/LC$), oscillations again occur. The charge on the condenser is

$$q = Q - \frac{Q}{\sqrt{1 - CR^2/4L}}\epsilon^{-(R/2L)t}\cos(\gamma t - \phi) \qquad (13\text{-}17)$$

$$\tan \phi = \frac{R/2L}{\sqrt{1/LC - R^2/4L^2}}$$

For small values of R, ϕ is nearly zero and Eq. (13-17) can be approximated by

$$q \simeq Q - Q\epsilon^{-(R/2L)t} \cos \left(\frac{1}{\sqrt{LC}} t \right)$$

Since R is small, the damping of the amplitude due to the exponential term will be slow. After approximately $\frac{1}{2}$ cycle of oscillation the cosine term will have the value -1; at this time $q \simeq 2Q$. The charge on the condenser is approximately twice the value which it could obtain from the battery alone (Fig. 13-12). Correspondingly, the voltage across the condenser is twice the voltage of the battery.

13-10. Voltage Developed across a Switch When a Circuit Is Broken. The previous work on transients will be applied to the problem of sparking at switches in circuits which contain inductance. It may be recalled that an amount of energy equal to $\frac{1}{2}LI^2$ joules is stored in a self inductance. When a circuit is broken, the current drops and this energy must be expended. It cannot remain stored as in the case of a capacitor in which the stored energy is a static thing, depending upon charge. In order to dissipate this stored energy, it is necessary for a current to flow for a finite time

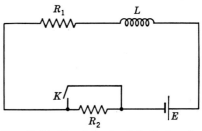

FIG. 13-13. Circuit to simulate the breaking of a circuit. The resistance R_2 is intended to represent the air-gap resistance when the key is opened.

after the key is opened; *i.e.*, a spark will occur at the key. Since the resistance of an ionized air gap is extremely complex, we shall simulate the conditions by means of the circuit illustrated in Fig. 13-13.

The circuit consists of a battery of voltage E, an inductance of L henrys, a resistance of R_1 ohms, and a key K. A second resistance R_2 is placed in parallel with the key and is to represent the resistance of the air gap when the key is opened.

The key is first closed and equilibrium is established as in Sec. 13-2. At a later time which shall be designated as $t = 0$, the key is opened, thereby suddenly increasing the resistance in the circuit. The problem is to find the voltage developed across the added resistance R_2. Applying Kirchhoff's law after the key is opened gives

$$L \frac{di}{dt} + i(R_1 + R_2) = E$$

Separating variables,

$$\frac{di}{E - i(R_1 + R_2)} = \frac{dt}{L}$$

This equation can be integrated directly:

$$\ln\left[E - i(R_1 + R_2)\right] = -\frac{R_1 + R_2}{L} t + K$$

The initial condition is at $t = 0$, $i = E/R_1$. Thus

$$K = \ln E \left(1 - \frac{R_1 + R_2}{R_1}\right)$$

Following the procedures of Sec. 13-2 gives the value

$$i = \frac{E}{R_1 + R_2}\left[1 + \frac{R_2}{R_1}\epsilon^{-(R_1+R_2/L)t}\right]$$

The voltage developed across the key is iR_2. Initially, at $t = 0$, this has the value

$$V_{t=0} = E\,\frac{R_2}{R_1}$$

In the actual case, an air gap replaces R_2, and this has essentially an infinite resistance initially. Thus the voltage across the gap will quickly build up to a point at which ionization of the gas will occur. The actual voltage which is initially developed across the added resistance R_2 is independent of the inductance in the circuit.

13-11. Elimination of Sparking at a Key in an Inductive Circuit.

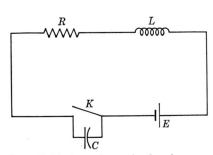

Sparking at an open key can be prevented if some method is devised for maintaining the voltage across the key at a low value. One such arrangement is illustrated in Fig. 13-14. A condenser is placed across the terminals of the key in the usual inductive circuit. Qualitative considerations are adequate to show the action of this circuit. Before the key is opened, all the stored energy resides in the inductance and is equal to $\frac{1}{2}LI^2$ joules.

FIG. 13-14. A condenser is placed across the key to "absorb" the energy stored in the inductance when the key is opened.

After the key has been opened, the current through the inductance will drop to zero and most of the energy will be stored in the capacitor. Thus, equating energies,

$$\tfrac{1}{2}LI^2 \simeq \tfrac{1}{2}CV^2$$

The maximum voltage across the capacitor and key is

$$V \simeq I\sqrt{\frac{L}{C}} = \frac{E}{R}\sqrt{\frac{L}{C}}$$

For example, suppose that E is 100 volts, R is 10 ohms, and L is 0.1 henry. If a 10-μf condenser were placed across the key, a maximum of 1,000 volts would be developed across the gap. In addition, this maximum voltage builds up in a finite time, thereby allowing the gap to become quite wide before the maximum voltage is reached.

TABLE 13-1. VALUES OF THE EXPONENTIAL ϵ^{-x}
Intermediate values can be obtained by interpolation.

x	ϵ^{-x}	x	ϵ^{-x}
0.00	1.000	1.10	0.333
0.05	0.951	1.20	0.301
0.10	0.905	1.30	0.273
0.15	0.861	1.40	0.247
0.20	0.819	1.50	0.223
0.25	0.779	1.60	0.202
0.30	0.741	1.70	0.183
0.35	0.705	1.80	0.165
0.40	0.670	1.90	0.150
0.45	0.638	2.00	0.135
0.50	0.607	2.50	0.0821
0.55	0.577	3.00	0.0498
0.60	0.549	3.50	0.0302
0.65	0.522	4.00	0.0183
0.70	0.497	4.50	0.0111
0.75	0.472	5.00	0.00674
0.80	0.450	5.50	0.00409
0.85	0.427	6.00	0.00248
0.90	0.407	6.50	0.00150
0.95	0.387	7.00	0.000912
1.00	0.368		

PROBLEMS

13-1. A capacitor C is charged from a battery E in series with a resistance R. Prove that the energy stored in the condenser when fully charged is equal to the energy lost in the resistance during the charging process.

13-2. A capacitor is discharged through the parallel circuit below. The total charge on the condenser is Q; the charge passing through R_g is Q_g. Prove that $Q_g/Q = R_s/(R_g + R_s)$. *Note:* This circuit can be considered to be a shunted ballistic

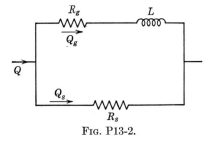

FIG. P13-2.

galvanometer having a coil resistance R_g and an inductance L, shunted by R_s. It should be noted that the division of charge is dependent only upon the resistance ratio.

13-3. The elements in the circuit of Fig. 13-1 have the values $E = 10$ volts, $R = 5$ ohms, $L = 0.1$ henry. Find the current in the circuit 0.01 sec after the key is closed. Find the current flowing after an elapsed time of four time constants.

13-4. Prove that L/R has the dimension of seconds.

13-5. Prove that RC has the dimension of seconds.

13-6. A 5-μf condenser is discharged through a 1-mh inductance in series with a 0.02-ohm resistance. What is the frequency of the alternating current during discharge? What is the magnitude of the charge at the end of the first, second, and third cycles of oscillation?

13-7. A battery of voltage E is connected in series with a key and a pure inductance. (Assume that there is zero resistance in the whole circuit.) Find the current flowing in the circuit as a function of time.

13-8. A 1-μf condenser is charged from a battery in series with a 10^4-ohm resistance. How much time will be required for the condenser to acquire 95 per cent of its ultimate charge?

13-9. If the charging battery in Prob. 13-8 has a voltage of 20 volts, find the current in the circuit 0.05 sec after the key is closed.

13-10. A capacitance is discharged through a resistance and inductance. The circuit is critically damped. If $R = 10$ ohms and $L = 1$ henry, find the fractional charge on the condenser after 0.01, 0.05, 0.1, and 1 sec.

CHAPTER 14

ALTERNATING CURRENT

14-1. Introduction. A current or voltage whose direction is a function of time is said to be alternating. The exact wave shape of such an alternating current or voltage may assume an infinite variety of forms. Some of the more common forms are illustrated in Fig. 14-1. It may appear, at first glance, that a thorough understanding of the action of such apparently diverse wave shapes is an insurmountable task. Fortunately it is

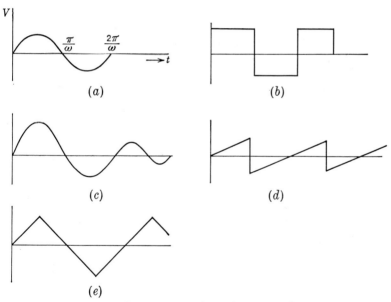

FIG. 14-1. Several alternating-voltage wave forms.

necessary to study only one of these in detail, namely, the sinusoidal form. This can be justified in several ways. First, voltages of the form $e = E_0 \sin \omega t$ are used nearly exclusively where electrical power is supplied commercially. This arises from the fact that a coil rotating in a uniform magnetic field cuts magnetic lines of flux at a rate proportional to $\sin \omega t$. Second, the common circuit elements which may be used in the electronic generation of alternating voltages naturally produce pure sine waves.   (This has been illustrated in the charging and discharging of a condenser through an inductance in Chap. 13.)   Third, when other wave forms are encountered, it is always possible to resolve them into a sum of pure sine waves by the application of Fourier's theorem.   We can thus reduce the study of alternating currents and voltages to one which involves only the sine-wave form without loss of generality.   In the following discussions we shall represent sources of voltage graphically by the symbol shown in Fig. 14-2 and in equation form by $e = E_0 \sin \omega t$. E_0 is the maximum value of this alternating voltage, ω is the angular frequency, $2\pi f$ or $2\pi/T$.   (f is expressed in cycles per second, while T is the period, or the time, required to complete 1 cycle.)

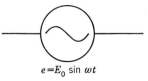

$e = E_0 \sin \omega t$

FIG. 14-2. Representation of a sinusoidal alternating-voltage source.

14-2. Effective, or Root-mean-square, Values of the Current and Voltage.   It is necessary to inquire as to the meaning of an a-c meter reading which gives the value of an alternating voltage or current as a specific numerical value.   When an alternating-voltage source is connected to a pure resistance of value R ohms, the current will vary sinusoidally ($i = I_0 \sin \omega t$) and electrical energy will be expended at the rate of $i^2 R$ joules/sec.   Clearly this power will vary continuously, and

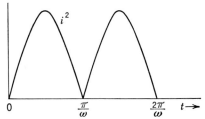

FIG. 14-3. Curve of the square of the alternating current as a function of time.

it will be more meaningful to speak of the average power developed in such a resistance.   This can be written (Fig. 14-3)

$$P_{\text{av}} = R(\text{av } i^2) = R\,\frac{\text{area under } i^2 \text{ curve}}{\text{base}}$$

$$= R \int^{\text{half cycle}} \frac{i^2\,dt}{\text{base}}$$

$$= R\,\frac{\int_0^{\pi/\omega} I_0{}^2 \sin^2 \omega t\,dt}{\pi/\omega} = R\omega I_0{}^2\,\frac{1}{2}\left(t - \frac{1}{2\omega}\sin 2\omega t\right)\Big|_0^{\pi/\omega}$$

$$= R\,\frac{I_0{}^2}{2} \tag{14-1}$$

For d-c circuits, the power consumed is $I^2 R$.   It is thus convenient to write the alternating power as

$$P = I_{eff}^2 R$$

in which

$$I_{eff} = \frac{I_0}{\sqrt{2}} \tag{14-2}$$

for a current which varies sinusoidally. For other wave shapes the effective current would bear a different relationship to the maximum value I_0. For example, the square wave illustrated in Fig. 14-1 has an effective value equal to the maximum value.

In the procedure above, two steps were followed to obtain the effective current. First, an average value of i^2 was obtained; second, the square root of this average value was taken so that $I_{eff} = \sqrt{\text{av } i^2}$. The name root-mean-square or simply rms value is another logical and commonly used name for the effective current.

In a similar manner the rms, or effective, voltage is found to be $E_0/\sqrt{2}$. These are the values indicated on normal a-c meters and voltmeters. Restated briefly, an alternating current which will produce the same quantity of heat per second as one ampere of direct current in a given resistance is said to have an effective, or rms, value of one ampere. Correspondingly, an alternating potential difference which produces the same heat per second in a given resistance as one d-c volt is said to have an effective value of one volt.

As an example, an a-c voltmeter placed across an ordinary lighting circuit might read 110 volts. The actual voltage will vary sinusoidally with time, having a maximum value of 155 volts ($110 \times \sqrt{2}$) but will produce the same amount of heat and light as would a direct voltage of 110 volts.

Some types of a-c meters can be calibrated with direct current and voltage by the techniques discussed previously. Such meters, *e.g.*, the electrodynamometer, are called transfer meters.

14-3. Simple Alternating Circuits.

An Alternating-voltage Source and a Pure Resistance. Figure 14-4 represents a simple series circuit consisting of a source of alternating voltage

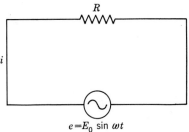

$e = E_0 \sin \omega t$

Fig. 14-4. A pure resistance connected across the terminals of an alternating-voltage source.

$e = E_0 \sin \omega t$ and a pure resistance of R ohms. Kirchhoff's second law is applicable at any instant of time, so that

$$iR = E_0 \sin \omega t$$

and
$$i = \frac{E_0}{R} \sin \omega t = I_0 \sin \omega t \qquad (14\text{-}3)$$

It is seen that the current through the resistance is in phase with the voltage across it (Fig. 14-5). That is, each varies as the sin ωt, and therefore the current will reach its maximum value E_0/R when the voltage has its maximum value of E_0.

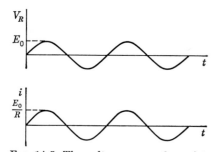

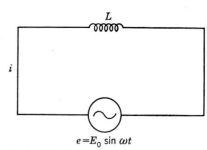

FIG. 14-5. The voltage across the resistance and the current through it as a function of time. It is noted that the current and voltage are in phase.

FIG. 14-6. A pure inductance connected to an alternating-voltage source.

An Alternating-voltage Source and a Pure Inductance. When the source of alternating voltage is placed across a *pure* inductance (Fig. 14-6), conditions are considerably different. Kirchhoff's law now is written

$$L \frac{di}{dt} = E_0 \sin \omega t$$

$$di = \frac{E_0}{L} \sin \omega t \, dt$$

$$i = -\frac{E_0}{\omega L} \cos \omega t = \frac{E_0}{\omega L} \sin (\omega t - 90°) \qquad (14\text{-}4)$$

The current is no longer in phase with the voltage but, in fact, lags behind the voltage by 90° (Fig. 14-7), so that when the voltage across the induct-

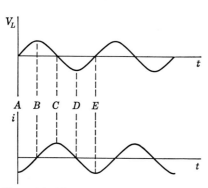

ance is at its maximum, the current in the circuit is rising through zero value. In addition, the maximum value of the current is $E_0/\omega L$, thus the quantity ωL, or $2\pi f L$, has the dimensions of resistance and plays the role of an effective resistance in limiting the current in an alternating circuit. This quantity is usually represented by the symbol $X_L = \omega L$ and is called the inductive reactance. For example, an inductance of 0.1 henry when used with an alternating source of 60 cycles per sec would have an inductive reactance of $X_L = 2\pi \times 60 \times 0.1 = 37.7$ ohms.

FIG. 14-7. The current is 90° behind the voltage for a pure inductance.

An Alternating-voltage Source and a Pure Capacitance. A simple capacitance circuit is shown in Fig. 14-8. The charge on the condenser at any instant of time is q, and the current flowing in the circuit is i.

Then

$$\frac{q}{C} = E_0 \sin \omega t$$

$$q = CE_0 \sin \omega t$$

$$i = \frac{dq}{dt} = \omega C E_0 \cos \omega t = \frac{E_0}{1/\omega C} \sin (\omega t + 90°) \qquad (14\text{-}5)$$

Again the current is out of phase with the voltage, in this case leading by 90° (Fig. 14-9). The maximum current is $\dfrac{E_0}{1/\omega C}$. The quantity $1/\omega C$ plays the role of an effective resistance in limiting the current and usually is given the symbol X_C, the capacitive reactance. It is, of course, measured in ohms.

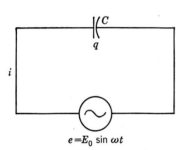

FIG. 14-8. A pure capacitance connected to an alternating-voltage source.

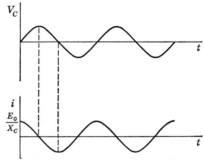

FIG. 14-9. The current leads the voltage by 90° for a pure capacitance.

14-4. Electrical Power Consumed in Simple Elements. *Resistance.* The power expended in an element at any instant of time is ei. For a pure resistance, the current and voltage are in phase, so that the average power is $1/T \int_0^T ei \, dt$ or $1/T \int_0^T (e^2/R) \, dt$. This has already been computed (Sec. 14-2) and found to be E_{eff}^2/R or $E_{eff}I_{eff}$ joules/sec. The electrical energy which is converted into heat in the resistance is $E_{eff}I_{eff}t$ joules, much like the d-c case.

Inductance. Referring to Fig. 14-7, during the time interval from A to B, voltage is positive while current is negative. The power, which is the product of the two, is therefore negative. For the next quarter cycle from B to C both are positive and the power is positive. Thus each quarter cycle the power changes from positive to negative, and the average over 1 cycle is zero. A positive value for the power means that energy is being supplied to the inductance from the voltage source. It is not dissipated but stored in the magnetic field of the inductance. A negative value for power represents the return of this stored energy to the source. No power is consumed in a pure inductance.

Capacitance. Again owing to the phase shift of 90° between the current and voltage, no energy is dissipated in a pure capacitance. Energy is alternately exchanged between the source and the condenser. Although inductive reactance and capacitive reactance play a role similar to a resistance in limiting an alternating current, they differ drastically in that power is not consumed in a reactance but may be dissipated in a resistance. In addition, a pure re-

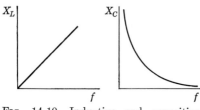

FIG. 14-10. Inductive and capacitive reactance as a function of frequency.

sistor has a resistance which is constant (at a given temperature) and "independent" of the frequency at which it is used. An inductive reactance increases with increasing frequency, and a capacitive reactance decreases with increasing frequency (Fig. 14-10).

14-5. Effective Values of the Current. The effective current is $I_{eff} = I_0/\sqrt{2}$, and the effective voltage is $E_{eff} = E_0/\sqrt{2}$. For an inductance $I_0 = \dfrac{E_0}{X_L}$ thus $I_{eff} = \dfrac{E_{eff}}{X_L}$. Similarly for a capacitor

$$I_{eff} = \frac{E_{eff}}{X_C}$$

An alternating source with frequency of 1,000 cycles per sec having an effective voltage of 50 volts is attached to a 10-μf condenser. An a-c ammeter connected in series with the condenser would read

$$I_{eff} =$$
$$\frac{50}{1/(2\pi \times 1,000 \times 10^{-5})} = 0.314 \text{ amp}$$

If the voltage of the source were held constant but the frequency changed to 10^4 cycles per sec, the effective current would rise to 3.14 amp.

14-6. Current in a Series Circuit. We shall apply the principles dis-

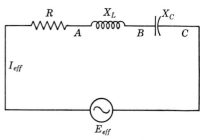

FIG. 14-11. A series a-c circuit. The influence of the frequency is taken into account in the values of X_L and X_C.

cussed above to a series circuit containing a resistance, inductance, capacitance, and alternating source of voltage (Fig. 14-11). The effective values of voltage and current will be used, since these are the quantities readily measured by a-c meters. The frequency of the source is, of course, important and is taken into account in computing the inductive and capacitive reactances X_L and X_C. Since we are dealing with a series circuit, the direction and magnitude of the current through each element are the same at any given instant. As a reference direction, let us take I along the positive x axis (Fig. 14-12a). The effective voltage across the resistance is in the direction of I (since the voltage across a resistance is in phase with the current through it) and has a magnitude $I_{eff}R$. The voltage across the inductance is $I_{eff}X_L$ and is plotted along the positive y axis (the current lags behind the voltage by 90° for an inductance). Similarly the voltage across the capacitance is $I_{eff}X_C$ plotted along the negative y axis. Adding these three voltages vectorially (Fig. 14-12b) gives the effective voltage of the source. For the sake of uniformity we can write the voltage of the source as $I_{eff}Z$, the current supplied by the source multiplied by an effective equivalent impedance measured in

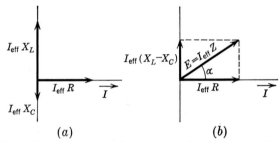

(a)   (b)

FIG. 14-12. "Vector" representation of the voltages in a series circuit. The reference direction is taken as the direction of the current.

ohms. In fact, this might be taken as the defining equation for the impedance for an alternating circuit

$$Z = \frac{E_{eff}(\text{source})}{I_{eff}(\text{source})} \qquad (14\text{-}6)$$

From the vector diagram (Fig. 14-12b) the value of Z for this series circuit is

$$Z = [R^2 + (X_L - X_C)^2]^{\frac{1}{2}} \qquad \text{ohms} \qquad (14\text{-}7)$$

The phase angle between the voltage of the source and the current supplied by it is the angle α, $\tan \alpha = (X_L - X_C)/R$.

Suppose that the voltage source in Fig. 14-11 is a commercial 60-cycle source with an effective voltage of 110 volts. Let $R = 20$ ohms, $L = 0.1$ henry, and $C = 100 \,\mu\text{f}$. Then $X_L = 2\pi \times 60 \times 0.1 = 37.7$ ohms and $X_C = 1/(2\pi \times 60 \times 10^{-4}) = 26.5$ ohms, $Z = [(20)^2 + (11.2)^2]^{\frac{1}{2}} = 22.9$, $I_{eff} = 110/22.9 = 4.8$ amp, $V_R = 4.8 \times 20 = 96$ volts,

$$V_L = 4.8 \times 37.7 = 181 \text{ volts}$$

$V_C = 4.8 \times 26.5 = 127.2$ volts. The current lags behind the voltage of the source by an angle α; $\tan \alpha = (37.7 - 26.5)/20 = 0.56$. A voltmeter connected from A to B would read 181 volts, from B to C 127.2 volts, but from A to C only 53.8 volts.

14-7. Power in an Alternating Circuit. The amount of power consumed in an alternating circuit may be investigated either from the standpoint of the power supplied by the source or by considering the power consumed in the individual circuit elements.

Referring to the circuit of Fig. 14-11 along with the vectorial representation of the voltages and current in Fig. 14-12, the source supplies a current I_{eff} at a voltage of E_{eff}. The voltage is out of phase with the current by an angle α. Let us break this voltage into two components: $E_{eff} \cos \alpha$, in phase with the current, and $E_{eff} \sin \alpha$, 90° out of phase with the current. The in-phase component will contribute energy at a rate $E_{eff}I_{eff} \cos \alpha$ joules/sec. The out-of-phase component will, on the average, contribute no power. Thus power can be written

$$P = E_{eff}I_{eff} \cos \alpha = I_{eff}^2 Z \cos \alpha \qquad (14\text{-}8)$$

$\cos \alpha$ is called the power factor for the circuit.

It has already been shown that no power is consumed in an inductance or capacitance. For a resistance, the current and voltage are always in phase; thus the power consumed is $I_{eff}^2 R$. If several resistances are present in a circuit, the total power is just the sum of the individual losses in the resistances. Thus

$$P = (I_{eff})_1^2 R_1 + (I_{eff})_2^2 R_2 \ldots \qquad (14\text{-}9)$$

For the example in Sec. 14-6,

$$P = E_{eff}I_{eff} \cos \alpha = 110 \times 4.8 \times 20/22.9 = 461 \text{ watts}$$

or $P = I_{eff}^2 R = 4.8^2 \times 20 = 461$ watts.

PROBLEMS

14-1. A coil is constructed which has an inductance of 10^{-3} henry and a resistance of 10 ohms. A voltage source of 5 volts, rms, is connected across the coil. Find the current through the coil, the phase angle between the current and voltage, and the power consumed for the following frequencies of the voltage: 0, 10^2, 10^3, 10^4, 10^5, and 10^6 cycles per sec.

14-2. The saw-tooth wave in Fig. 14-1d has a maximum value of E_0 volts and a period of T sec. Calculate the effective voltage.

14-3. Show that ωL and $1/\omega C$ have the same dimensions as resistance.

14-4. Using the circuit of Figs. 14-6 and 14-7, prove that the energy consumed is zero by evaluating the integral $W = \int_0^t ei \, dt$.

14-5. In the circuit below, find the expression for the voltage across the inductance and the voltage across the resistance. How much power is being dissipated in the resistance? What would be the reading of an ammeter placed in this circuit?

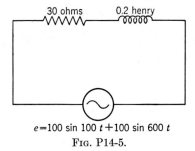

$e = 100 \sin 100\, t + 100 \sin 600\, t$

FIG. P14-5.

14-6. The effective current ($\sqrt{\text{average } i^2}$) has been shown to be $0.7071E_0$. Show that the average current is $(2/\pi)\, E_0$ or $0.64E_0$ if the average is taken over $\frac{1}{2}$ cycle.

CHAPTER 15

SOLUTION OF A-C CIRCUITS BY MEANS OF
COMPLEX NOTATION

15-1. Introduction. In the previous chapter it has been shown that, for a simple alternating series circuit, a phase difference may exist between the voltages across various elements. In this sense voltages must be treated as vector quantities having magnitude and direction (with respect to each other or with respect to the current flowing). Kirchhoff's second law must be treated in vector form; that is, the vector sum of all the voltages in a completed circuit must equal zero.

For a parallel section of an alternating circuit, the instantaneous voltage must be the same for each branch at all times. This can, in general, occur only if a phase difference exists between the currents in the various branches. Thus it is necessary to treat the currents as vector quantities, and Kirchhoff's first law can be stated as follows: At any branch point the vector sum of the currents entering the branch point equals the vector sum of the currents leaving.

It should be obvious that a straightforward solution of an alternating circuit can become extremely tedious, since one must take into account both the magnitude of a quantity and its direction for each part of the circuit. Fortunately, there are two procedures which can be used to aid in the solution of such circuits. The graphical procedure can be used, wherein the vectors representing the various quantities are plotted in the proper direction and with appropriate magnitudes. The results can then be obtained from the measurement of a length and an angle on the graph. While this procedure is quite instructive and gives the student a certain "feeling" for the quantities involved, it is not adaptable to problems requiring good precision.

A second method is that of complex notation wherein impedances are expressed in the form of complex numbers to aid in the calculations and manipulations. Once the various procedures have been carried out, the artificially introduced complex form is eliminated and quickly yields voltages, currents, and relative phases in all parts of the circuit. It may be looked upon as a tool for simplifying the solution of a-c circuits.

15-2. Complex Numbers. The vector A in Fig. 15-1 is plotted along the positive x direction. When A is multiplied by the number (-1) its magnitude remains the same but its direction has been rotated through an angle of 180°. Multiplication by (-1) a second time returns the vector to its original position by producing an additional rotation of 180°. Since multiplication of a vector by $(-1)^2$ produces a rotation of the vector through 360° and multiplication by $\sqrt{(-1)^2}$ produces a rotation through half the angle (180°), it seems reasonable to assume that multiplication of the vector by $\sqrt{-1}$ would produce a rotation of 90°. This unit vector $\sqrt{-1}$ will be called j.

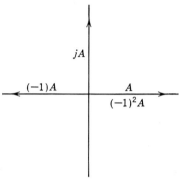

FIG. 15-1. The vector A multiplied by the "operators" 1, (-1), $(-1)^2$, and $(-1)^{1/2}$ or j.

The quantity $A + jB$ is a complex number; A is called the real part; jB is the imaginary part. It is unfortunate that the word "imaginary" has a rather indefinite and almost mysterious connotation to many. The complex number should be considered as a convenient method for representing a vector and for carrying out certain operations. Figure 15-2a shows a vector which represents a force of 5 lb at an angle of 37° with the x axis. A second, equivalent method for describing this same force (Fig. 15-2b) is by means of its components along the x and y axes. That is, the statement $F_x = 4$ lb and $F_y = 3$ lb contains the same information as the original representation of the vector. A third method for representing the 5-lb force is as follows: $4 + j3$. This complex number merely represents the two components of

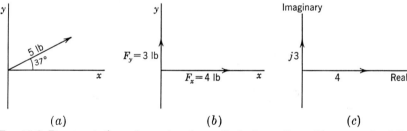

(a)   (b)   (c)

FIG. 15-2. Representations of a vector of magnitude five units making an angle of 37° with the x axis.

the original 5-lb vector, 4 lb along the real or x axis and 3 lb along the imaginary or y axis.

15-3. Complex Notation Applied to Series Circuits. In general, the rules of algebra are followed in dealing with complex numbers. The number $A_1 + jB_1$ when added to $A_2 + jB_2$ yields $(A_1 + A_2) + j(B_1 + B_2)$.

Similarly the difference between two vectors

$$(A_1 + jB_1) - (A_2 + jB_2) = (A_1 - A_2) + j(B_1 - B_2)$$

The unit vector j, $\sqrt{-1}$, when multiplied by itself equals -1. Thus $(A_1 + jB_1)(A_2 + jB_2) = (A_1A_2 - B_1B_2) + j(B_1A_2 + A_1B_2)$. The mag-nitude of the vector represented by the complex number $A + jB$ (Fig. 15-3) is $r = \sqrt{A^2 + B^2}$; its direction with respect to the real axis is $\theta = \tan^{-1} B/A$. For the complex number $A_1 + jB_1$ to be equal to $A_2 + jB_2$ two conditions must be ful-filled: $A_1 = A_2$ and $B_1 = B_2$.

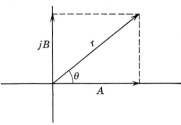

FIG. 15-3. The magnitude of the vector $|r| = \sqrt{A^2 + B^2}$ while $\theta = \tan^{-1} B/A$.

15-4. Complex Notation Applied to Series Circuits. A simple series cir-cuit has been discussed in the preceding chapter. It has been seen that the various voltages (IR, IX_L, and IX_C) have certain phases or directions with respect to the current. Since the current through each element is the same in magnitude and direction, we can divide each term by the cur-rent, thus leaving a vector diagram for the impedances (Fig. 15-4). The individual components can be written in complex notation as follows: $R, jX_L,$ and $-jX_C$. The resultant impedance of the series circuit written

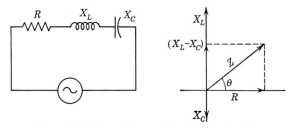

FIG. 15-4. Phase diagram for a series circuit.

in complex form is $Z = R + j(X_L - X_C)$. The magnitude of the imped-ance is $|Z| = \sqrt{R^2 + (X_L - X_C)^2}$; its direction with respect to the real axis is $\theta = \tan^{-1} (X_L - X_C)/R$. (The effective voltage of the whole circuit *leads* the current by the angle θ.)

15-5. Complex Notation Applied to Parallel Circuits. The voltage drop across the individual branches of a parallel circuit is the same, and this voltage must be identical with the voltage across any equivalent ele-ments which replace the parallel branch. (This is the case whether one refers to the instantaneous, maximum, or effective voltage.) In addition, the vector sum of the currents through all branches must equal the cur-rent flowing through the equivalent element which replaces a parallel branch.

Figure 15-5a is a simple parallel circuit. I_R, I_L, and I_C represent the effective currents through the resistance, inductance, and capacitance, respectively. Z_{eq} in Fig. 15-5b is the single equivalent impedance which can replace the parallel branch.

We shall work with admittances in the elementary treatment rather than with impedances. An admittance Y is the reciprocal of an impedance, so that $Y_R = 1/R$, $Y_L = 1/X_L$, $Y_C = 1/X_C$, and $Y_{eq} = 1/Z_{eq}$.

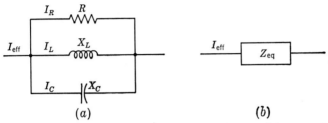

FIG. 15-5. A parallel circuit and the equivalent impedance which can replace it.

The various phase relationships are indicated in Fig. 15-6a. The reference direction is the direction of the voltage, since this is common to all elements, and for convenience it is set along the x axis. The current through the resistance is $I_R = VY_R$, and its direction is the same as V, since for a pure resistance, voltage and current are always in phase. The current through the capacitance $I_C = VY_C$ and is along the positive y axis, since the current through a capacitance leads the voltage by 90°. Similarly I_L is along the negative y axis, lagging behind the voltage by 90°.

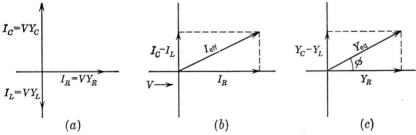

FIG. 15-6. Phase diagram for the currents in a parallel circuit. The voltage serves as the reference direction.

The vector sum of the currents I_{eff} is indicated in Fig. 15-6b. If each current is divided by V, an admittance diagram is obtained (Fig. 15-6c). The equivalent admittance $Y_{eq} = \sqrt{Y_R^2 + (Y_C - Y_L)^2}$. The current in the equivalent circuit *leads* the voltage by an angle

$$\phi = \tan^{-1} \frac{(Y_C - Y_L)}{Y_R}$$

It is noted that the vector treatment of parallel circuits using admittances is very similar to the treatment of series circuits using impedances. It is, therefore, reasonable to expect that one would gain the same advantages in treating parallel circuits by means of complex notation as were gained with series circuits. Following this procedure we can write the complex equivalent admittance $\mathbf{Y}_{eq} = Y_R + jY_C - jY_L$. Since admittance and impedance are reciprocal, $1/\mathbf{Z}_{eq} = (1/R) + (j/X_C) - (j/X_L)$. If the numerator and denominator of the last two terms are multiplied by j, we have $1/\mathbf{Z}_{eq} = (1/R) + (1/-jX_C) + (1/jX_L)$; that is, the reciprocal law applies to parallel impedances using the usual complex notation (that is, $-jX_C$ and jX_L).

15-6. Rules for the Solution of Circuits by Complex Notation. We shall attempt to set down a group of rules or procedures which may aid the beginning student in the solution of alternating circuits by the procedures of complex notation.

1. Write all impedances in complex form. This may be done by inspection.

Impedance	Complex form
R	R
X_L	jX_L
X_C	$-jX_C$

2. Using the complex impedances, carry out all manipulations in the manner of d-c circuits; that is, series impedances add in complex form, parallel impedances obey the reciprocal law in complex form, etc.

3. Reduce all complex impedances to the form $A + jB$. In the normal manipulation it often happens that the complex impedance is in the form $(E + jF)/(H + jK)$. If both numerator and denominator are multiplied by the complex conjugate of the denominator, in this case $H - jK$, the denominator becomes a real number and the complex impedance is in the form of $A + jB$.

4. After the whole circuit has been reduced to a single complex impedance $\mathbf{Z} = A + jB$, the magnitude of the impedance can be found: $Z_{real} = \sqrt{A^2 + B^2}$.

5. The effective current supplied by the source $I_{eff} = V_{eff}/Z_{real}$.

6. The voltage leads the current by an angle $\theta = \tan^{-1} B/A$.

The circuit in Fig. 15-7 will be solved to illustrate these procedures.

Impedance	Complex form
$R_1 = 20$ ohms	20
$X_{L_1} = 15$ ohms	$j15$
$R_2 = 10$ ohms	10
$X_{C_1} = 10$ ohms	$-j10$
$X_{L_2} = 20$ ohms	$j20$
$R_3 = 5$ ohms	5
$X_{C_2} = 8$ ohms	$-j8$

$$\mathbf{Z}_{AC} = 20 + j15$$

$$\mathbf{Z}_{CE} = \frac{(10 - j10)j20}{(10 - j10) + j20} = \frac{20 + j20}{1 + j} \times \frac{1 - j}{1 - j} = \frac{20 + 20}{2} = 20 \text{ ohms}$$

$$\mathbf{Z}_{EF} = \frac{5(-j8)}{5 - j8} = \frac{-j40}{5 - j8} \times \frac{5 + j8}{5 + j8} = \frac{320 - 200j}{25 + 64} = \frac{320}{89} - \frac{200}{89}j$$
$$= 3.6 - 2.25j$$

$$\mathbf{Z}_{AF} = \mathbf{Z}_{AC} + \mathbf{Z}_{CE} + \mathbf{Z}_{EF} = 43.6 + 12.75j$$

This shows immediately that the whole circuit from A to F could be replaced by a resistance of 43.6 ohms in series with an inductive reactance of 12.75 ohms. It may also be noted, from the equivalent complex

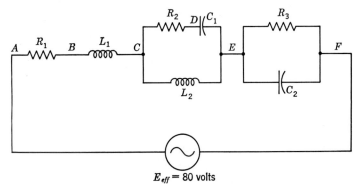

$E_{eff} = 80$ volts

FIG. 15-7. An a-c circuit.

impedance $\mathbf{Z}_{CE}$, that the parallel combination from C to E can be replaced by a pure resistance of 20 ohms. The magnitude of the actual impedance of the whole circuit is

$$|\mathbf{Z}_{AF}| = \sqrt{43.6^2 + 12.75^2} = \sqrt{1{,}920 + 162} = \sqrt{2{,}082} = 45.6 \text{ ohms}$$

The effective current $I_{eff} = 80/45.6 = 1.75$ amp

$$V_{AB} = 1.75 \times 20 = 35 \text{ volts}$$
$$V_{BC} = 1.75 \times 15 = 26.25 \text{ volts}$$
$$V_{CE} = 1.75 \times |\mathbf{Z}_{CE}| = 1.75 \times 20 = 35 \text{ volts}$$
$$V_{EF} = 1.75 \times |\mathbf{Z}_{EF}| = 1.75 \sqrt{3.6^2 + 2.25^2} = 1.75 \sqrt{13 + 5.1}$$
$$= 1.75 \times 4.25 = 7.45 \text{ volts}$$

The current in the lower branch between C and E is

$$I_{lower} = \frac{V}{X_{L_2}} = \frac{35}{20} = 1.75 \text{ amp}$$

In the upper branch between C and E the current is

$$I_{upper} = \frac{35}{|\mathbf{Z}_{upper}|} = \frac{35}{\sqrt{10^2 + 10^2}} = \frac{35}{14.1} = 2.48 \text{ amp}$$

The voltage across R_2 in the upper branch is

$$V_{R_2} = 2.48 \times 10 = 24.8 \text{ volts}$$
and
$$V_{C_1} = 2.48 \times 10 = 24.8 \text{ volts}$$

The current through R_3 is $I_{R_3} = 7.45/5 = 1.5$ amp. The current through C_2 is $I_{C_2} = 7.45/8 = 0.93$ amp.

The phase angle between the current and the voltage for the circuit as a whole can be obtained from the complex impedance of the whole circuit $\mathbf{Z}_{AF} = 43.6 + 12.75j$, thus $\tan \theta = 12.75/43.6$ and $\theta = 16°$. Similarly the phase angle between the current and voltage for the parallel branch between C and E is obtained from the complex impedance between C and E, $\mathbf{Z}_{CE} = 20 + 0j$. Thus $\tan \theta_{CE} = 0/20$ and the current flowing into this parallel branch is in phase with the voltage across it. For the parallel branch from E to F the phase angle is θ_2 and $\tan \theta_2 = -2.25/3.6$. $\theta_2 = -32°$. (The negative sign indicates that the current leads the voltage.) The power can be obtained from the over-all circuit:

$$P = E_{eff}I_{eff} \cos \theta = 80 \times 1.75 \times \frac{43.6}{45.6} = 134 \text{ watts}$$

The power can also be found by considering the losses in the resistance.

$$P = \Sigma I^2 R = (1.75)^2 \times 20 + (2.48)^2 \times 10 + (1.5)^2 \times 5$$
$$= 3.1 \times 20 + 6.2 \times 10 + 2.25 \times 5$$
$$= 62 + 62 + 11.3 = 135 \text{ watts}$$

(The difference between the values of the power as obtained by the two procedures arises from rounding off numbers in the calculations.)

15-7. Series Resonance. The series circuit (Fig. 15-4) has already been discussed. The complex impedance of this circuit is

$$\mathbf{Z} = R + j(X_L - X_C)$$

The current is $I_{eff} = E_{eff}/\sqrt{R^2 + (X_L - X_C)^2}$. The values of both X_L and X_C are frequency dependent, X_L increasing with an increase in f while X_C decreases with an increase in f. At one frequency $X_L = X_C$, and for this value, $f = 1/2\pi \sqrt{LC}$, the impedance of the circuit is a minimum equal to R, and the circuit is said to be resonant. The current is E_{eff}/R, and the current supplied by the source is in phase with the voltage of the source.

Illustration. Let $E_{eff} = 10$ volts, $R = 2$ ohms, $L = 0.1$ henry, and $C = 10^{-7}$ farad. This circuit would be in resonance for a frequency of $10^4/2\pi$ cycles per sec. At this frequency $X_L = 10^3$ ohms and

$$X_C = 10^3 \text{ ohms}$$

The current in the circuit is 5 amp. The voltage across the resistance is 10 volts, across the inductance 5,000 volts ($I_{eff}X_L$), and across the capacitance 5,000 volts.   The voltage developed across the inductance (or capacitance) is 500 times as large as the source voltage.   Such a series resonant circuit thus provides a means of amplifying a voltage at the resonant frequency.   Circuits of this type are often used to supply relatively high voltages as a substitute for a transformer when little power is required from the high-voltage source.   A plot of the impedance as a function of frequency is shown in Fig. 15-8.   At frequencies below the resonant frequency f_0, the circuit acts like a resistance-inductance combination with the current lagging behind the voltage by an angle $\tan^{-1}(X_L - X_C)/R$.

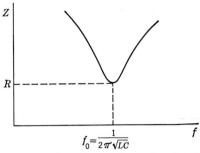

FIG. 15-8. Impedance of a series circuit as a function of frequency.   At resonance the impedance is equal to R.

At frequencies higher than f_0, capacitance effects predominate and the current leads the voltage.

15-8. Parallel Resonance or Antiresonance.   The complex impedance of the parallel circuit (Fig. 15-9), is

$$Z = \frac{(R + jX_L)(-jX_C)}{R + j(X_L - X_C)}$$

$$= \frac{RX_C^2}{R^2 + (X_L - X_C)^2} - \frac{jX_C(R^2 + X_L^2 - X_L X_C)}{R^2 + (X_L - X_C)^2}$$

At resonance, $X_L = X_C$ and

$$Z = \frac{X_C^2}{R} - jX_C$$

The magnitude of the impedance $|Z| = X_C\sqrt{(X_C^2/R^2) + 1}$.   The impedance of the circuit for this resonant frequency, $f = 1/2\pi\sqrt{LC}$, is a maximum (Fig. 15-10).   In fact, for $R = 0$ the impedance of an antiresonant circuit is infinite.

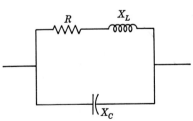

FIG. 15-9. A parallel circuit.

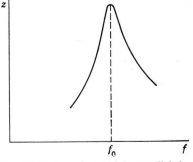

FIG. 15-10. Impedance of the parallel circuit as a function of frequency.   The impedance is a maximum at resonance.

15-9. Filters. An electrical filter is nothing more than a potential divider which contains one or more elements having an impedance which is dependent upon frequency. This gives rise to a condition whereby the voltage, as measured across some element, will be a function of both the

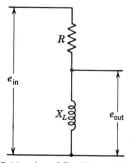

FIG. 15-11. An *LR* filter. The low-frequency components will be suppressed in the output.

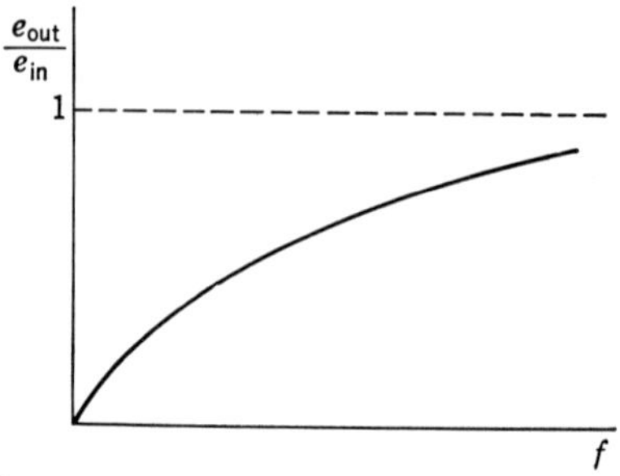

FIG. 15-12. Voltage ratio for an *LR* filter.

total applied voltage and the frequency. Figure 15-11 is such a filter, consisting of a resistance in series with an inductance. The input or applied voltage is impressed across the two elements in series; the output voltage is taken across the inductance. (It is assumed that no current is drawn at the output leads.) The current supplied by the source is $I = e_{in}/\sqrt{R^2 + X_L^2}$; the voltage drop across the inductance is

$$e_{out} = IX_L = e_{in}\frac{X_L}{\sqrt{R^2 + X_L^2}}$$

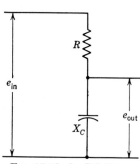

FIG. 15-13. An *RC* filter.

Since $X_L = 2\pi fL$, the output voltage of this potential divider will be a function of frequency, the ratio e_{out}/e_{in} approaching unity at high frequencies (Fig. 15-12). Such a filter would "pass" the higher frequencies while suppressing low frequencies.

A second type of filter is shown in Fig. 15-13. The ratio of the output to the impressed voltage is $e_{out}/e_{in} = X_C/\sqrt{R^2 + X_C^2}$. This quantity approaches unity for low frequencies while suppressing the higher frequencies (Fig. 15-14).

The series and parallel resonance circuits discussed previously can be used as elements of filters which transmit or reject narrow bands of frequencies.

Using the circuits of Figs. 15-11 and 15-13, measure the ratio of e_{out}/e_{in} over a wide range of frequencies and for several values of the resistances

R. The output voltage should be measured with a high-impedance meter such as a vacuum-tube voltmeter. For the RL filter the resistance of the inductance coil itself must be taken into consideration (Fig. 15-15).

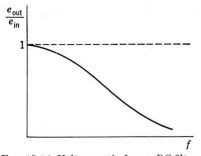

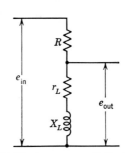

FIG. 15-14. Voltage ratio for an RC filter. The higher frequencies are suppressed in the output.

FIG. 15-15. A more realistic LR filter. r_L represents the coil resistance.

15-10. Inductors in Series. A practical coil which is used as an inductance inherently consists of a resistance r_L, the ohmic resistance of the wire, in series with the inductance of the coil itself. Figure 15-16a represents two such inductance coils in series. The methods of complex notation will be used to find the equivalent impedance of this combination:

$$Z = (r_{L_1} + r_{L_2}) + j(\omega L_1 + \omega L_2)$$

It is evident that the two coils are equivalent to a resistance $r_L = r_{L_1} + r_{L_2}$ in series with an inductance $L = L_1 + L_2$ (Fig. 15-16b).

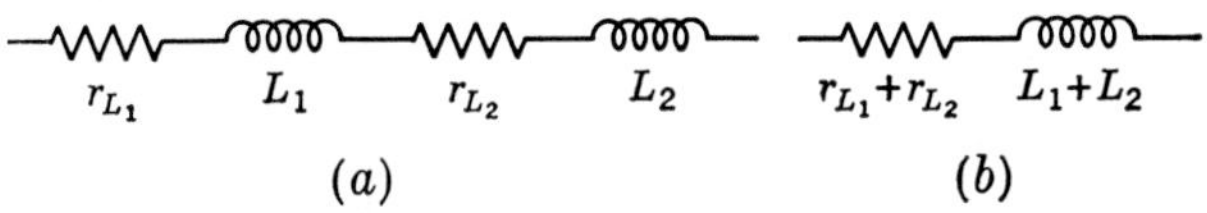

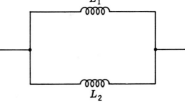

$$(a) \qquad\qquad (b)$$

FIG. 15-16. (a) Two inductance coils in series. (b) Equivalent circuit for the two inductance coils.

15-11. Inductances in Parallel. Figure 15-17 represents two pure inductances connected in parallel. The complex impedance of the combination is

$$Z = \frac{(j\omega L_1)(j\omega L_2)}{j(\omega L_1 + \omega L_2)}$$

$$= j\omega\left(\frac{L_1 L_2}{L_1 + L_2}\right) = j\omega L_{eq}$$

FIG. 15-17. Two inductances in parallel.

That is, two pure inductances can be replaced by a single equivalent inductance L_{eq} which has a value equal to the product of the individual inductances divided by their sum; i.e., the reciprocal law applies.

PROBLEMS

15-1. In the accompanying circuit, find the current through each element, the voltage drop across each, the phase angle between the current and the voltage for the whole circuit, and the power consumed.

a. $R = 5$ ohms   *b.* $R = 5$ ohms
    $X_L = 6$ ohms $X_L = 6$ ohms
    $X_C = 12$ ohms $X_C = 6$ ohms

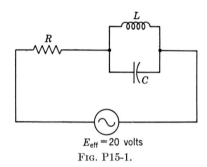

$E_{\text{eff}} = 20$ volts
FIG. P15-1.

15-2. *a.* Find the impedance of the parallel circuit below at a frequency of 60 cycles per sec.
    b. At what frequency does resonance occur?
    c. What is the impedance at resonance?

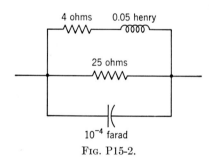

FIG. P15-2.

15-3. Making use of the procedures of complex notation prove that:
    a. Several capacitors C_1, C_2, C_3, . . . in parallel can be replaced by a single capacitor C such that $C = C_1 + C_2 + C_3 + \cdots$
    b. Several capacitors in series can be replaced by a single capacitor according to the reciprocal law $1/C = (1/C_1) + (1/C_2) + \cdots$
15-4. Two coils are placed in series. One has a time constant of 0.04 sec and a resistance of 20 ohms. The second has a time constant of 0.1 sec and a resistance of 10 ohms. Find the time constant of the series combination.
15-5. The two coils in Prob. 15-4 are placed in parallel. Find the time constant.
15-6. Draw the wye circuit which is equivalent to the accompanying delta circuit.

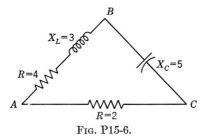

Fig. P15-6.

15-7. A source of voltage consisting of three frequency components is impressed on an *RC* filter. Find the voltage across the capacitance for each of the frequency components indicated in the diagram.

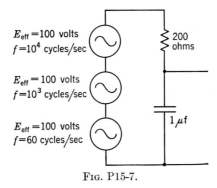

Fig. P15-7.

15-8. The source of voltage in Prob. 15-7 is connected to an *RL* filter. Find the voltage across the inductance for each frequency component.

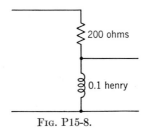

Fig. P15-8.

15-9. In the circuit of Probs. 15-7 and 15-8, a resistive load of 2,000 ohms is placed across the condenser or inductance. Find the voltage across these elements for each frequency component.

15-10. At times a single filter will not suppress an undesired frequency sufficiently. It is then possible to apply the "filtered" voltage to a second filter in order to reject further the undesired components. In the accompanying diagram, find the output voltage for each component of frequency and compare this with the single filter of Prob. 15-7.

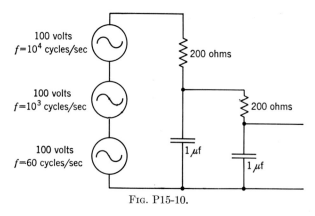

FIG. P15-10.

15-11. A certain measurement is to be made at a frequency of 60 cycles per sec obtained from the commercial power source. It is found, however, that this source also contains a 120-cycle component. This is to be eliminated by means of the

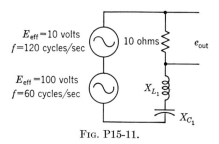

FIG. P15-11.

filter in the accompanying circuit. Compute the voltage for each frequency at the output of this filter system.

$$X_{L_1}\begin{cases} 1{,}000 \text{ ohms at 60 cycles} \\ 2{,}000 \text{ ohms at 120 cycles} \end{cases}$$
$$X_{C_1}\begin{cases} 1{,}000 \text{ ohms at 60 cycles} \\ 500 \text{ ohms at 120 cycles} \end{cases}$$

15-12. A resistor having pure ohmic resistance cannot be constructed. The resistive material and leads will always contain a certain amount of inductance and capacitance which may play an important role at high frequencies. The equivalent circuit for such a resistor is indicated in the accompanying diagram. For a good resistor with $R = 100$ ohms, L might typically be 10^{-6} henry and $C = 10^{-11}$ farad.

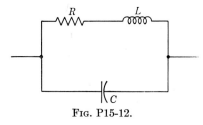

FIG. P15-12.

a. Calculate the impedance when used at 10^4, 10^6, and 10^8 cycles per sec.

b. Find the resonant frequency, and calculate the impedance at this frequency.

c. What is the power factor at 10^4, 10^6, and 10^8 cycles per sec?

CHAPTER 16

MEASUREMENT OF CAPACITANCE

16-1. Introduction. The measurement of capacitance by means of the ballistic galvanometer has already been discussed, and its inherent difficulties and limitations have been noted. The bridge method, using alternating currents, provides a null procedure for comparing impedances. These bridge methods usually are more satisfactory from the standpoint of convenience, flexibility, and accuracy. Figure 16-1 is an a-c bridge of the general Wheatstone type. Z_1, Z_2, Z_3, and Z_4 represent various impedances. When the bridge is balanced, the potential difference between A and B is at all times zero. Thus $I_A Z_1 = I_B Z_3$ and

$$I_A Z_2 = I_B Z_4$$

When one of these equations is divided by the other, the currents are eliminated and we have the general expression for the a-c bridge,

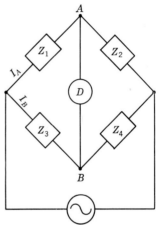

FIG. 16-1. Impedance bridge.

$$\frac{Z_1}{Z_2} = \frac{Z_3}{Z_4} \quad \text{or} \quad Z_1 Z_4 = Z_2 Z_3 \quad (16\text{-}1)$$

In order that these two complex numbers shall be equal to each other, two conditions of balance must be fulfilled, their real parts must be equal, and their complex parts must be equal. For example, if $Z_1 = R_1 + jX_1$, $Z_2 = R_2 + jX_2$, $Z_3 = R_3 + jX_3$, and $Z_4 = R_4 + jX_4$, then

$$(R_1 + jX_1)(R_4 + jX_4) = (R_2 + jX_2)(R_3 + jX_3)$$
$$(R_1R_4 - X_1X_4) + j(R_4X_1 + R_1X_4) = (R_2R_3 - X_2X_3)$$
$$+ j(R_2X_3 + R_3X_2)$$

The bridge is balanced if

$$R_1R_4 - X_1X_4 = R_2R_3 - X_2X_3$$

and
$$R_4X_1 + R_1X_4 = R_2X_3 + R_3X_2$$

16-2. Detectors for the A-C Bridge. The ordinary telephone receiver or headset is one of the most common and satisfactory detectors for use

239

with the a-c bridge when the frequency of the source is in the audible region.   A good headset is able to detect an unbalance of the order of 1 mv or a current of the order of 1 μamp.   This is usually capable of giving a balance point with an accuracy compatible with the precision of the various components of the bridge.

If a bridge is to be used at frequencies above the audible range, an oscilloscope may be useful as a detector.   The amplitude of the trace is a measure of the amount of unbalance in the bridge.   When higher sensitivity is required, an electronic amplifier can be used to increase the signal before it is applied to the headset or oscilloscope.

In addition to the headset and oscilloscope it is possible to use a d-c galvanometer as a detector by inserting a crystal diode in series with the galvanometer.   The voltage sensitivity of the galvanometer is drastically reduced from its d-c value when operated in this fashion to detect small

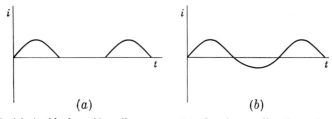

(a)   (b)

FIG. 16-2. (a) An ideal rectifier allows current to flow in one direction only.   (b) At low voltage, the rectifier is only partly effective.

potential differences.   This arises from two effects.   First, the diode resistance is in series with the galvanometer coil and this resistance is quite high at low applied voltages; second, the rectifying action of a diode becomes less effective at low voltages (Fig. 16-2); i.e., the difference between the forward and reverse resistance is small.   To illustrate these points, a d-c galvanometer with coil resistance of 10^3 ohms and current sensitivity of 3×10^{-8} amp/cm was used with a 1N34 diode.   The normal voltage sensitivity is 3×10^{-5} volt/cm.   Using the diode arrangement with direct current the voltage sensitivity was reduced to 7×10^{-4} volt/cm due to the added diode resistance.   With an alternating current the sensitivity was further reduced to 7×10^{-3} volt/cm.   When such sensitivities are adequate it may be advisable to use this type of detector.

TRANSIENT METHODS

16-3. Bridge Method for Comparing Two Capacitances.   This is a null method and therefore capable of more exact measurements than the direct-deflection method of the ballistic galvanometer.   The two condensers are placed in two arms of the Wheatstone bridge setup as shown in Fig. 16-3. The galvanometer G is a ballistic galvanometer measuring the total charge which passes through it. Its resistance is r_g. When the key is depressed, current will start to flow through the resistors to charge the condensers. Applying Kirchhoff's second law to the loop ABF gives

$$i_1 R_1 + i_g r_g - i_2 R_2 = 0 \qquad (16\text{-}2)$$

Since $i = dq/dt$, this equation can be written

$$R_1 \frac{dq_1}{dt} + r_g \frac{dq_g}{dt} - R_2 \frac{dq_2}{dt} = 0$$

Initially, at $t = 0$, the charge $q = 0$. Integration yields

$$R_1 Q_1 + r_g Q_g - R_2 Q_2 = 0$$

The values, capital Qs, represent the total charge which has passed through the resistors during the charging process. If the bridge is balanced, the galvanometer produces no deflection when the key is closed; therefore $Q_g = 0$ and $R_1 Q_1 = R_2 Q_2$. After the condensers have become fully charged, the current stops flowing and the voltage across each is the same; that is, $Q_1/C_1 = Q_2/C_2$. Combining these results and eliminating the quantities Q_1 and Q_2 gives

$$R_1 C_1 = R_2 C_2 \qquad (16\text{-}3)$$

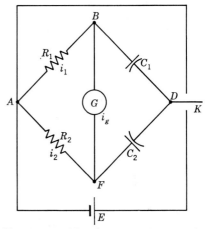

FIG. 16-3. Bridge for measuring capacitance using transient currents.

This condition of balance holds equally well for the discharging process when the key is raised to the upper contact.

It is interesting to note that an inductance in either or both of the resistance arms of the bridge will not influence the conditions of balance. This is so since an inductance in a given arm, say in series with R_1, would produce an additional term in Eq. (16-2), namely, $L(di/dt)$. The current is initially zero and also finally zero; the integration of this additional term yields $Li \Big|_0^0 = 0$ and contributes nothing to the over-all process.

(The presence of such inductance may, indeed, cause charges to flow through the ballistic galvanometer in one direction during part of the charging process, but it will cause an equal and opposite charge during another part. The resulting effect will still be zero deflection of the ballistic galvanometer if its period is long compared with the charging process.)

16-4. Comparison of Capacitances by the Series Capacitance Bridge.
Gott's Method.   This is another bridge method and differs in arrangement
from the preceding only by having the galvanometer and battery interchanged.   In the bridge method (Sec. 16-3) the balance is obtained for
the conditions which exist *during* the charging or the discharging of the
condensers.   In the present method the capacitances of the condensers
are compared after everything has reached the steady and permanent
condition.   Thus if one (or both) of the condensers has resistance in series
with itself, this will not affect the value of the final charge in the con-

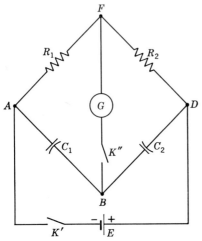

denser, and the capacitance is readily
measured.

     The arrangement is shown in Fig.
16-4.   When the battery key K' is
closed, the two condensers in series
are charged to the difference of
potential between A and D.   The
point B, between the two condensers,
has a potential intermediate between
that of A and D.   The measurement
consists in adjusting R_1 and R_2 until
F, on the upper circuit, has the same
potential as B in precisely the same
way as in the measurement of resistance by the Wheatstone bridge
method.   When this adjustment is
correct, there will be no deflection of
the galvanometer upon closing K''.

Fig. 16-4. Series capacitance bridge.

Since the condensers are joined in series, each one must contain the
same charge.   The point B is insulated as long as K'' remains open, and
therefore whatever charge appears on one condenser must have come from
the other one.   Moreover, for a balance, the closing of K'' produces no
deflection of the galvanometer; *i.e.*, there is no change in the charges on
the condensers joined at B.

     Writing out the equations for the potential differences in the circuits
$BAFB$ and $DBFD$ gives

$$\frac{q}{C_1} - R_1 i = 0 \quad \text{and} \quad \frac{q}{C_2} - R_2 i = 0$$

from which
$$C_1 = C_2 \frac{R_2}{R_1}$$

     It should be noted that closing K'' brings B to the potential of F,
whether there is a balance or not.   A second closing of K'' can produce
no deflection.   It is therefore necessary to discharge the condensers

completely after each closing of K''. This can be most quickly done by opening K' before opening K''.

If one of the condensers has considerable absorption or leakage, it will seriously influence the results, for after some of the charge has leaked away, it is no longer true to say that the charges on the two condensers are equal. The effect of this source of error is reduced by closing K'' as soon as possible after closing K'. When the resistances R_1 and R_2 are free from inductance and capacitance, it is allowable to omit K'' from Fig. 16-4 and to observe the deflections of the galvanometer when K' is closed or opened.

16-5. Comparison of Capacitances by the Method of Mixtures. This method was devised by Lord Kelvin to avoid some of the difficulties in the preceding methods. It is especially applicable to cases where the two capacitances are very dissimilar, e.g., if the capacitance of a long cable is to be compared with that of a stand-ard condenser. The method con-sists in charging the condensers to such potentials that each will con-tain the same quantity of charge. They are then discharged, the one into the other, and the charges al-lowed to mix. If the charges are not equal, the difference will remain in the condensers and is later dis-charged through the galvanometer.

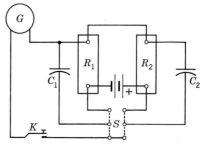

Fig. 16-5. Method of mixtures.

The arrangements and connections are shown in Fig. 16-5. Two resist-ance boxes of moderately high resistance, R_1 and R_2, are joined in series with a battery of a few cells. A small current i is allowed to flow continu-ously through R_1 and R_2, thus maintaining across R_1 a steady potential difference,

$$V_1 = R_1 i$$

and across R_2 a potential difference,

$$V_2 = R_2 i$$

The condenser to be measured is connected in parallel with R_1 through the double-pole, double-throw switch S. It is thus charged to the poten-tial difference of $R_1 i$ volts and therefore receives a charge

$$Q_1 = C_1 R_1 i$$

At the same time, the known condenser is connected in parallel with R_2 by the other side of the switch. The charge it receives is

$$Q_2 = C_2 R_2 i$$

By opening S, these condensers are disconnected from R_1 and R_2, but each condenser still retains its charge. When S is closed on the other side, each of the condensers discharges into the other, and the electrons constituting the negative charge of one condenser thus "mix" with the positive charge of the other condenser. If the charges in the two condensers are equal, one will just neutralize the other, and there will be no resultant charge in either condenser. In case the charge in one condenser is larger than that in the other, a part of this charge will be left after the "mixing," part of it being in each condenser.

Closing the tapping key K enables the condensers to be discharged through the ballistic galvanometer. If there is zero deflection, it shows that there was no charge remaining after the mixing, and therefore

$$Q_1 = Q_2$$

This condition can be obtained by varying R_1, and therefore Q_1, until zero deflection is obtained. Then

$$C_1 R_1 i = C_2 R_2 i$$

or

$$C_1 = C_2 \frac{R_2}{R_1}$$

ALTERNATING-CURRENT METHODS

16-6. Wien Capacitance Bridge. An a-c bridge for comparing capacitances is shown in Fig. 16-6. The resistances r_1 and r_2 in series with the capacitances represent the equivalent series resistance of the condenser. While the series resistance would be zero for an ideal condenser, power is always expended in a real capacitance. The power losses can be lumped together as though occurring in a single equivalent resistance r such that $P = I^2 r$. For the bridge to be balanced

$$\frac{R_1}{r_1 - (j/\omega C_1)} = \frac{R_2}{r_2 - (j/\omega C_2)}$$

$$R_1 r_2 - j \frac{R_1}{\omega C_2} = R_2 r_1 - j \frac{R_2}{\omega C_1}$$

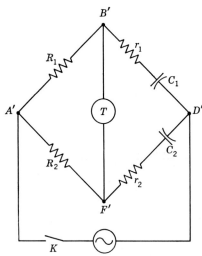

Fig. 16-6. Alternating-current capacitance bridge.

or $R_2/R_1 = r_2/r_1$ (real parts equal) and $R_2/R_1 = C_1/C_2$ (imaginary parts equal).

It is usually not possible to satisfy both of these conditions simultaneously with the circuit illustrated. This is indicated by the lack of complete silence in the earphone. While it is often satisfactory to accept the minimum sound in the detector as the balance point, a more complete balance can be obtained by inserting a small resistance in series with r_1 or r_2 so that both equations of balance will be satisfied. This added resistance is usually of the order of 1 ohm or less.

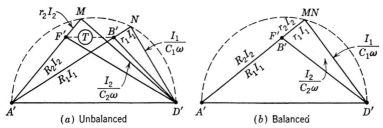

(a) Unbalanced   (b) Balanced

FIG. 16-7. Vector method for solving the bridge circuit in Fig. 16-6.

16-7. Grounding the Bridge. When a current flows through a bridge, the various parts of the apparatus are brought to different potentials. If these parts have any electrostatic capacitance between themselves and the surroundings, it will require the addition of corresponding charges thus to change the potentials. If any of these charges flow in through the telephone receivers, perhaps to charge the receivers themselves, there will not be silence at the correct balance point. Since the observer is usually connected to the earth, the parts of the bridge connected to the telephones should be at zero potential also. If F' (Fig. 16-8), were grounded, the potential of the telephone would remain at zero, owing to the addition of sufficient charges at this point, but such additional charges are not desirable in the bridge.

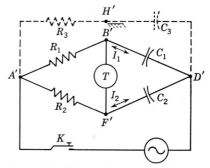

FIG. 16-8. Illustration of the Wagner ground.

The same effect can be obtained by joining another circuit, $A'H'D'$, in parallel with the bridge, as shown by the dotted part of Fig. 16-8. This circuit is made similar to the bridge by setting $R_3 = R_1C_1/C_3$ with the aid of the telephone receiver connected temporarily between B' and H' after the bridge is approximately balanced. By grounding the point H', the points B' and F' are kept at zero potential without the addition of extra charges at these points.

Other bridges can be grounded in this manner[1] by a circuit of resistance, inductance, and capacitance similar to the bridge itself. In case it is desired to keep some other point of the bridge at zero potential, the corresponding point on the auxiliary circuit is grounded.

16-8. Series Capacitance Bridge with Alternating Current. A second method for comparing capacitances with an a-c bridge is shown in Fig. 16-9. Again r_1 and r_2 represent the effective resistances of the condensers. At balance

$$\frac{R_1}{R_2} = \frac{r_1 - (j/\omega C_1)}{r_2 - (j/\omega C_2)}$$

or $R_2/R_1 = r_2/r_1$ and $R_2/R_1 = C_1/C_2$.

The best balance is obtained by inserting a small resistance in series with either r_1 or r_2 and by the use of a Wagner ground, which in this case will consist of two resistances in the ratio of R_1 to R_2.

16-9. Impedance Bridge. An impedance bridge capable of measuring both the resistive and reactive

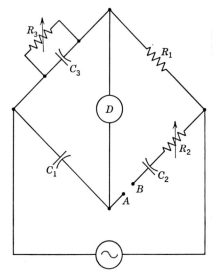

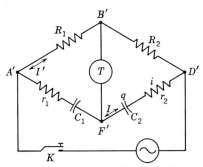

Fig. 16-9. Series capacitance bridge.

Fig. 6-10. Impedance bridge. Resistive and reactive components can be measured simultaneously.

components of an impedance over a wide range of values is shown in Fig. 16-10. The bridge is first balanced with the junctions A-B shorted. The equation is

$$\frac{-jX_{C_3}R_3/(R_3 - jX_{C_3})}{R_1} = \frac{-jX_{C_1}}{R_2 - jX_{C_2}}$$

$$-jR_2R_3X_{C_3} - R_3X_{C_3}X_{C_2} = -jR_1R_3X_{C_1} - R_1X_{C_1}X_{C_3}$$

For balance

$$R_3X_{C_2} = R_1X_{C_1} \quad \text{and} \quad R_2X_{C_3} = R_1X_{C_1}$$

[1] This is a Wagner ground.

After this "zero" balance is obtained, the unknown impedance $R + jX$ is inserted between the terminals A-B and a new balance is obtained by readjusting the variable resistances to R'_2 and R'_3. The equation is now

$$\frac{-jX_{C_3}R'_3/(R'_3 - jX_{C_3})}{R_1} = \frac{-jX_{C_1}}{R'_2 + R - j(X_{C_2} - X)}$$

$$-jX_{C_3}R'_2R'_3 - jX_{C_3}R'_3R - R'_3X_{C_3}X_{C_2} + R'_3X_{C_3}X = -jX_{C_1}R_1R'_3$$
$$- X_{C_3}X_{C_1}R_1$$

$$R = \frac{R_1X_{C_1} - R'_2X_{C_3}}{X_{C_3}}$$

but from the "zero" balance $R_1X_{C_1} = R_2X_{C_3}$. Thus $R = R_2 - R'_2$, the difference between the zero balance value of R_2 and its value with the unknown impedance in place. Similarly

$$X = \frac{X_{C_2}R'_3 - X_{C_1}R_1}{R'_3}$$

Again from the "zero" balance conditions $X_{C_1}R_1 = R_3X_{C_2}$ and

$$X = X_{C_2}\left(1 - \frac{R_3}{R'_3}\right)$$
$$= \frac{1}{\omega C_2}\left(1 - \frac{R_3}{R'_3}\right)$$

The resistive component of the impedance is obtained from the adjustment of the series resistance R_2 and is independent of the source frequency. The reactive component X is obtained from the setting of R_3, and although it is dependent upon frequency, the dials can be calibrated for direct reading if the bridge is used at a fixed frequency, say 1,000 cycles per sec. A positive value for X indicates an inductance, while a negative value is obtained for a capacitive reactance.

16-10. Power Factor of a Condenser by a Resistance-shunted Bridge Method. When power is absorbed in one of the condensers that are being compared by the bridge method, the condenser current, instead of being in exact quadrature with the applied voltage, becomes slightly more nearly in phase with it. This means that the currents in the two resistance arms of the bridge are not quite in phase with each other. This makes it impossible to obtain an exact balance by merely changing the ratio of the resistance arms.

The phase of the current in the other branch of the bridge can be changed by adding a high resistance W (Fig. 16-11) in parallel with the standard condenser. By varying the resistance in W, the power expended in the resistance-condenser combination can be made to correspond with that expended in the unknown condenser. This changes the phase of I_2 to equal that of I_1, and the bridge can be balanced.

The vector diagram corresponding to the bridge of Fig. 16-11 is shown in Fig. 16-12. Varying the resistance W will change the value of i and therefore the angle φ'. Changing R_2 will change the relative lengths of AF and FD. By varying first one and then the other, the point F can be brought to coincide with B, and the bridge is balanced. The angle φ now equals the angle φ'.

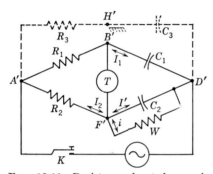

FIG. 16-11. Resistance-shunted capacitance bridge.

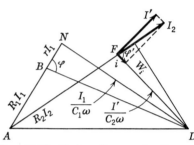

FIG. 16-12. Vector diagram for the unbalanced resistance-shunted bridge.

From the figure, the power factor of the unknown condenser C_1 is

$$\cos \varphi = \cos \varphi' = \frac{i}{I_2} = \frac{i}{I'} = \frac{E/W}{EC_2\omega} = \frac{1}{WC_2\omega}$$

since I_2 is very nearly the same as I' for this large value of W.

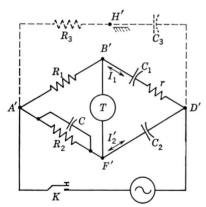

FIG. 16-13. Capacitance-shunted bridge.

With this small value of $\cos \varphi$ the line ND is practically as long as BD, and so we also have

$$C_1 = C_2 \frac{R_2}{R_1}$$

as in the previous case.

16-11. Power Factor of a Condenser by a Capacitance-shunted Bridge Method. If there is available a variable condenser of small capacitance, e.g., from zero to 0.001 μf, it is very convenient to use it as a shunt on one of the resistance arms of the capacitance bridge, as shown in Fig. 16-13. This will add another component to the current in C_2, and by adjusting this component, i, the emf across C_2 can be brought into phase with that across the condenser (r, C_1).

The vector diagram for the balanced condition is shown in Fig. 16-14. r represents the equivalent series resistance within the condenser C_1.

When the standard condenser C_2 has no power loss and $C = 0$, the diagram AFD for the lower branch of the bridge coincides with AND, and the bridge is not balanced. By increasing C (Fig. 16-13) a component i is added to I_2 to make up the total current I_2' in the standard condenser C_2. This will throw I_2' out of phase with I_1 and will bring F nearer to B. When F and B coincide, as shown in the diagram, the bridge is balanced.

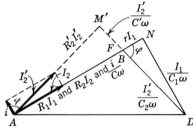

FIG. 16-14. Diagram for the balanced condition of the capacitance-shunted bridge.

Observe that at A, i and I_2 combine into I_2', thus making the branch $A'F'$ behave like a series circuit of R_2' and C', as shown by the dashed lines of the diagram.

The power factor of condenser C_1 is

$$\cos DBN = \cos \varphi = \cot \varphi = \frac{i}{I_2} = \frac{EC\omega}{E/R_2} = R_2C\omega$$

and this can be determined as exactly as C is known.

16-12. Shielded Circuits. In every bridge arrangement there is a small amount of capacitance between its conductors and the surroundings—the walls of the room, the hands of the operator, etc. A varying potential applied to these conductors means a varying charge and therefore extra currents through parts of the bridge. These extra currents cause a poor and false balance of the bridge. Their effect is somewhat reduced by the use of a Wagner ground, which keeps the potential of the telephone or other detector at the potential of the surrounding walls and thus reduces the charges on this part of the bridge.

To reduce still further the effect of this variable capacitance, each arm of the bridge and each connecting conductor should be individually enclosed within a well-grounded metal shield. This does not eliminate the capacitance to the surroundings (shields), but it makes it constant, and its effect is counted as a part of the circuit when that is measured and calibrated.

16-13. Substitution. In measurements by the bridge method, the ratio between the known and unknown values is determined. When these values are nearly equal, it is often better to make one balance with the unknown condenser in the bridge and then a second balance with a known condenser standing in the place of the unknown, with the other arms of the bridge undisturbed.

For the first balance,

$$C_1 = C_2 \frac{R_2}{R_1} \tag{16-4}$$

For the second balance,

$$C_s = C_2 \frac{R_2}{R_1 + R} \tag{16-5}$$

where R is the resistance that must be added to R_1 for the second balance.
Dividing (16-4) by (16-5) gives

$$C_1 = C_s \left(1 + \frac{R}{R_1}\right) \tag{16-6}$$

where all uncertainty in the actual values of C_2 and R_2 disappears and the unchanged part of R_1 enters only in the small fraction R/R_1.

PROBLEMS

16-1. In the accompanying circuit, derive the conditions for balance. Show how one can obtain the power factor for the branch A-B.

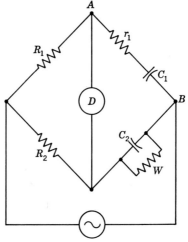

Fig. P16-1.

16-2. A real condenser can be represented as in the diagram. If $C = 100$ $\mu\mu$f, $R_S = 1$ ohm, and $R_P = 10^{12}$ ohms, find the impedance between A and B at the following frequencies: 1, 10^4, 10^6, and 10^{10} cycles per sec. Also find the impedance of an ideal condenser at these frequencies (that is, $R_S = 0$ and $R_P = \infty$).

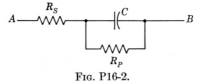

Fig. P16-2.

CHAPTER 17

MEASUREMENT OF SELF INDUCTANCE, MUTUAL INDUCTANCE, AND FREQUENCY

17-1. Introduction. Many of the basic techniques which have already been discussed can be applied in the measurement of inductance and frequency. In many cases the most convenient procedure involves the use of the a-c bridge. When this is used, the usual precautions in grounding the bridge and choosing the proper detector should be observed.

SELF INDUCTANCE

17-2. Comparison of Self Inductance, Series Method. Figure 17-1 indicates a bridge arrangement for comparing a self inductance L' with another self inductance L and several noninductive resistances. R and S represent resistances in the inductive arms of the bridge and include both the resistance of the coil and that of a resistance box included in the arm. At balance, the usual conditions exist:

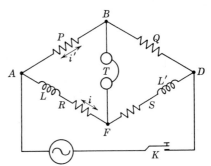

$$i'P = iR + ijX_L$$
and
$$i'Q = iS + ijX_{L'}$$
$$\frac{P}{Q} = \frac{R + jX_L}{S + jX_{L'}}$$
$$PS + jPX_{L'} = QR + jQX_L$$

Fig. 17-1. Bridge for comparing self inductance.

Since both the real and complex parts must be equal at balance,

$$PS = QR \quad \text{or} \quad \frac{P}{Q} = \frac{R}{S}$$

and
$$PX_{L'} = QX_L \quad \text{or} \quad \frac{L}{L'} = \frac{P}{Q}$$

The process by which this double adjustment can be made is as follows: If the inductances of the coils are fixed values L and L', the resistance P can be set at about 1,000 ohms; Q is varied to give a first approximation

251

to a minimum sound in the telephone. Probably this will be far from silence, because R and S are not set at the proper values. If one of these, say S, is varied, the sound can be reduced to a lower minimum. Now it will be possible to readjust Q to a sharper minimum, and then S can be readjusted. By successively varying Q and S, the correct value for each can soon be reached. If perfect silence is not attained, it is probably because of a lack of fineness in the adjustment of the resistances.

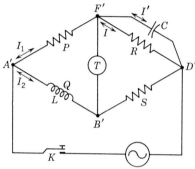

FIG. 17-2. Comparison of a self induct-ance with a capacitance.

If the known self inductance is also variable, a balance can be obtained by adjusting this variable self induct-ance and one of the resistances.

17-3. Measurement of Self In-ductance by Comparison with a Ca-pacitance, Maxwell's Method. In this method the self inductance to be measured is placed in one arm of a bridge, the other arms of which should be as free from inductance as possible. Balance can be obtained by placing a condenser in parallel with a resistance R as shown in Fig. 17-2. At balance

$$I_1 P = I_2(Q + jX_L) \tag{17-1}$$

$$I_1 \frac{R(-jX_C)}{R - jX_C} = I_2 S \tag{17-2}$$

Dividing Eq. (17-1) by (17-2) yields

$$\frac{P}{-jRX_C/(R - jX_C)} = \frac{Q + jX_L}{S}$$

or $$PSR - jPSX_C = -jQRX_C + RX_CX_L$$

The conditions of balance are

$$PS = QR$$

and $$PS = \frac{L}{C}$$

In practice, a capacitance is chosen which has a value approximately equal to L/RQ. The final balance is then obtained by alternately adjust-ing the resistances P and R.

17-4. Measurement of Self Inductance by Comparison with a Capaci-tance, Anderson's Method. A variation of Maxwell's method is indi-cated in Fig. 17-3. In this procedure, one additional resistance r is con-nected in series with the capacitor and one end of the detector is attached

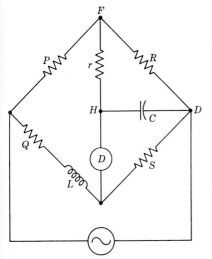

FIG. 17-3. Anderson's method for comparing a self inductance with a capacitance.

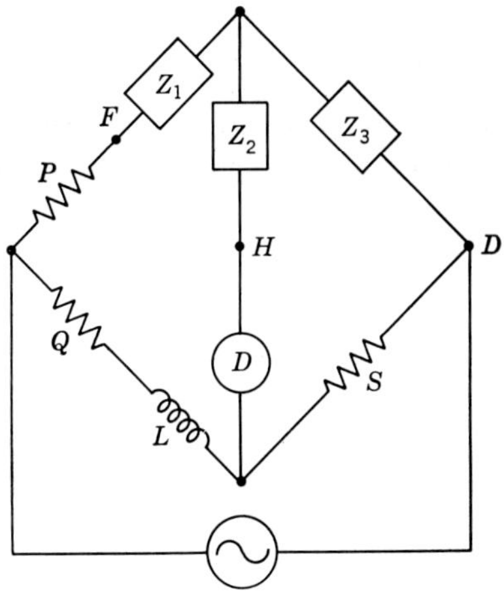

FIG. 17-4. The delta in Anderson's circuit has been replaced by an equivalent wye made of Z_1, Z_2, and Z_3.

at the junction of r and C. This circuit can be solved by applying a Δ-Y transformation, converting the Δ (made of r, R, and C) into the Y (made up of Z_1, Z_2, and Z_3), as shown in Fig. 17-4. It is evident that

$$Z_1 = \frac{rR}{r + R - jX_C}$$

$$Z_2 = -\frac{jX_C r}{r + R - jX_C}$$

and
$$Z_3 = -\frac{jX_C R}{r + R - jX_C}$$

When the bridge is balanced, the current through the detector, and thus through Z_2, is zero.

The usual bridge equations can be written:

$$\frac{P + rR/(r + R - jX_C)}{-jX_C R/(r + R - jX_C)} = \frac{Q + jX_L}{S}$$

or
$$PSr + PSR + rRS - jX_C PS = -jX_C RQ + RX_L X_C$$

Equating the complex parts yields the first condition of balance:

$$PS = QR \qquad (17\text{-}3)$$

The real parts give the second condition of balance:

$$L = PSC + rC\,\frac{PS}{R} + rCS$$

$$= PSC + rC(Q + S) \qquad (17\text{-}4)$$

It is noted that both conditions of balance are independent of frequency and somewhat independent of each other. For example, it is possible to obtain the first condition of balance [Eq. (17-3)], using a d-c source of voltage and a galvanometer for a detector. After this has been carried out, an a-c source and telephone can be inserted, and by varying r and C, Eq. (17-4) can be satisfied without disturbing the balance obtained with direct current.

MUTUAL INDUCTANCE

17-5. The Emf of Mutual Inductance. When there is an alternating current in one of the coils of a mutual inductance, there is an emf in the other coil. The value of this emf can be found from the following considerations.

Let the instantaneous value of the alternating current in the primary coil be written as

$$i = I' \sin \omega t$$

The flux turns through the secondary coil due to this current will be

$$(\Phi n) = Mi$$

where M denotes the value of the mutual inductance between the coils. The emf in the secondary coil is measured by the rate of change in this flux through it and is

$$e = \frac{d(\Phi n)}{dt} = M \frac{di}{dt}$$

or, using the value of i,

$$e = M \frac{di}{dt} = M\omega I' \cos \omega t$$

The maximum value of this emf is $M\omega I'$, where I' is the maximum value of the primary current. The effective value is

$$E \text{ (effective)} = M\omega I$$

where I is the effective value of the primary current.

In any circuit that may be connected to the secondary coil, this emf will produce a current in the same way as any other alternating emf. In most cases it is not necessary to refer to its relation to the primary current, except in its value of $M\omega I$. *In a vector diagram containing both I and $M\omega I$, the vector $M\omega I$ would be drawn $90°$ ahead of I, the same as with $L\omega I$ in a similar diagram.*

17-6. Direct Comparison of Two Mutual Inductances. Mutual inductances are readily measured by comparing the emfs induced in their secondaries. If the current through the primary coils is direct, the induced emf will appear at the make and break of the circuit; while if alternating current is used, there will be an alternating emf in the secondary.

In the direct comparison of two mutual inductances (Fig. 17-5) the two primaries are joined in series and connected to the source of current, preferably a low-voltage alternating circuit of audio frequency. The two secondaries are also joined in series, with the two induced emfs opposed to each other.

When the same current flows through the primary coils, the emfs induced in the secondaries are proportional to the mutual inductances of the coils. If the two secondaries are joined in series with a telephone, or other current detector, the effective value of the secondary current is

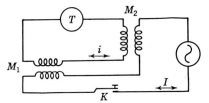

FIG. 17-5. Comparison of two mutual inductances that can be adjusted to equality.

$$i = \frac{M_1 \omega I - M_2 \omega I}{Z}$$

where Z is the total impedance of the secondary circuit. If, now, the mutual inductance of one of the coils can be varied until the telephone shows by its silence that $i = 0$, we have

$$M_1 = M_2$$

This method requires a variable standard of mutual inductance, and a telephone to indicate zero current in the secondary circuit. By turning the movable coil of the standard until the emf induced in it is just equal and opposite to that induced in the secondary of the other pair of coils, as shown by zero sound in the telephone, the value of the mutual inductance can be read directly from the standard.

It is instructive to interchange the primary and secondary coils of one of the mutual inductances. It will be found that the value of the mutual inductance of this pair of coils is the same, whichever coil is used for the primary.

17-7. Compensation for the Impurity of a Mutual Inductance. In the diagram of Fig. 17-5, it is assumed that $M_1 \omega I$ and $M_2 \omega I$ are the only emfs in the secondary circuit and that at the balance point there should be absolute silence in the telephones in the secondary circuit. However, this is not always the case.

There is a certain amount of capacitance between the conductors that form the primary and secondary windings of a coil, and when there is a difference of potential between these conductors, a capacitance-current component is added to the primary current. This means that the primary currents in the two coils are not strictly in phase with each other, because a part of the current is used to charge this capacitance. Therefore $M_1\omega I_1$ is not exactly in opposite phase with $M_2\omega I_2$ and cannot entirely balance it.

In this case a better balance can be obtained by introducing into the secondary circuit a small emf which will differ in phase from $M_2\omega I_2$ by 90°. Such an emf is found across a small resistance in the primary circuit, as shown by r in Fig. 17-6. This resistance may be only a short length of wire in the primary circuit, depending upon the amount of the capacitance between the two windings. Joining the primary and secondary circuits as shown in Fig. 17-6 helps to reduce any large difference of potential between them.

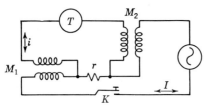

FIG. 17-6. Compensation for the impurity of a mutual inductance.

17-8. Comparison of Two Unequal Mutual Inductances. *With Direct Current and a Ballistic Galvanometer.* Let M be a pair of coils whose mutual inductance is known and M' another pair whose mutual inductance is desired. The primaries of the two coils are joined in series with a battery and key, the coils themselves being placed as far apart as possible and at right angles to each other, so that each secondary will be influenced only by its own primary. The two secondaries are also joined in series with two resistance boxes, and a galvanometer is connected across from B to C. This is not a Wheatstone bridge arrangement, for the two emfs in the two secondaries act together and the current flows through the four arms in series.

By properly adjusting R_1 and R_2, the galvanometer will give no deflection when K is opened or closed.

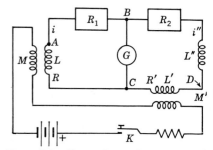

FIG. 17-7. Comparison of two mutual inductances.

Writing the instantaneous values of the potential differences for the circuit $CABC$ gives, at the instant t,

$$M \frac{dI}{dt} - Ri - L \frac{di}{dt} - R_1 i - Gi' - L_g \frac{di'}{dt} = 0$$

and for $DCBD$

$$M' \frac{dI}{dt} - R'i'' - L' \frac{di''}{dt} + Gi' + L_g \frac{di'}{dt} - R_2 i'' - L'' \frac{di''}{dt} = 0$$

where L_g denotes the inductance of the galvanometer circuit.

Integrating each of these equations from the time when the key is first closed and $I = 0$, to the time when the primary current has reached its steady value $I = I_0$, gives

$$(R + R_1)q = MI_0 \quad \text{and} \quad (R' + R_2)q'' = M'I_0$$

when q', the integrated current through the galvanometer, is zero. Dividing one equation by the other gives

$$\frac{M}{M'} = \frac{R + R_1}{R' + R_2}$$

since $q = q''$, when $q' = 0$.

In this method the galvanometer current may be zero, but in general it consists of two transient currents, each of which has an effect upon the galvanometer even when following one another in a short interval of time. This may produce an unsteadiness of the galvanometer and render an exact setting difficult, if not impossible. In order that no current should pass through the galvanometer, it is necessary that the potential difference between B and C shall remain zero for each instant while the primary current is changing, and this requires that the self inductance of each branch shall be proportional to the emf induced in that part of the circuit (see Fig. 17-7).

Usually this is not the case, but it is not difficult to add some self inductance L'' in the part of the circuit that is deficient and thus fulfill this condition. The galvanometer will then indicate a much closer balance, or it may be replaced by a telephone, with an alternating current used in the primaries as shown in Sec. 17-9.

17-9. Comparison of Two Unequal Mutual Inductances. *With Alternating Current and a Telephone.* When it is not possible to balance one mutual inductance against the other, as in the method of Sec. 17-6, they can be compared by the arrangement shown in Fig. 17-8. The two secondaries are joined in series, helping each other, and connected to an external circuit consisting of two resistance boxes in series. The current through this circuit is due to the combined emfs of both coils. This may also be thought of as one coil supplying the emf needed to keep the current flowing through its part of the

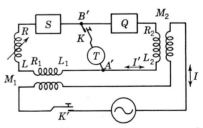

Fig. 17-8. Comparison of two unequal mutual inductances. R_1 is the resistance in the secondary coil L_1. R_2 denotes the resistance in the secondary coil L_2.

circuit and the other emf being used to keep the current flowing through the rest of the circuit. If these portions of the circuit can be measured, the relative values of the emfs are determined.

17-10. The Vector Diagrams. Since the two primaries are in series and carry the same current, the emfs induced in their secondaries will be in phase with each other. When K is open, there is only a single current through the secondary circuit. The emf diagram for the secondary circuit is then as shown in Fig. 17-9a.

Starting at O, the vector $(S + R + R_1)I'$ is laid off horizontally to the right, where $(S + R + R_1)$ denotes the total resistance from B' to A' through the secondary coil L_1. Since $(L + L_1)\omega I'$ is 90° ahead of $(S + R + R_1)I'$, it is drawn as FA.

Likewise $(Q + R_2)I'$ is drawn parallel to OF as AD, where $(Q + R_2)$ denotes the total resistance in the secondary coil L_2 and the added resistance box Q. Then $L_2\omega I'$ is drawn as DP. These four vectors

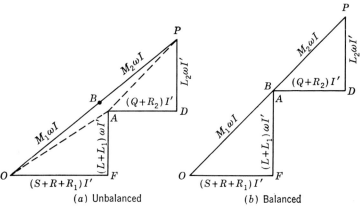

(a) Unbalanced   (b) Balanced

Fig. 17-9. Emf diagrams for two mutual inductances in series.

from O to P show the components of the emf that is required to maintain the current I' in the combined secondary circuit of the two coils. The emf available for this purpose is $M_1\omega I + M_2\omega I$. Therefore this is also drawn as OP.

The diagram for the unbalance case shows that $M_1\omega I$ is not quite enough, either in phase or in amount, to maintain the current in its part of the circuit, lacking by the voltage shown as AB. The telephone will respond to this voltage. When S and L are adjusted to give zero sound in the phones, $M_1\omega I$ is just enough to maintain the current I' in its part of the circuit from B around to A, while $M_2\omega I$ will maintain the current through the rest of the circuit.

By varying S and L, the point A can be brought to coincide with B. For this case no current will flow through a telephone joined from A to B, and therefore this diagram (Fig. 17-9b) corresponds to the arrangement when there is silence in the telephone. The two triangles are similar and give the relation

$$\frac{S + R + R_1}{Q + R_2} = \frac{M_1}{M_2} = \frac{L_1 + L}{L_2}$$

If S and Q are very large resistances, a good balance can be obtained even when L_1 and L_2 are neglected, but in arranging the circuit it is best to fulfill this relation as nearly as possible. For the best balance a variable self inductance should be included in the secondary circuit and adjusted as may be found necessary to reduce the sound from the telephone.

17-11. Comparison of Two Unequal Mutual Inductances. *Divided-potential Method.* A simple and convenient arrangement for comparing mutual inductances is shown in Fig. 17-10. The two primaries are in series, and therefore each carries the same current. The emf induced in the secondary coil of the smaller mutual inductance M_1 is balanced against a part of the emf from the secondary coil of the larger mutual inductance M_2. This balance is not true at each instant, but it is an integrated balance and is shown by a ballistic galvanometer. One of the advantages of this method is that the resistance of the secondary coil in the smaller mutual inductance need not be known.

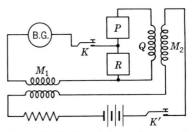

FIG. 17-10. Comparison of unequal mutual inductances by the divided-potential method.

When the equations of potential differences in the two secondary circuits have been integrated, the ratio of the two mutual inductances is found to be

$$\frac{M_1}{M_2} = \frac{R}{P + Q + R}$$

where Q is the resistance of the secondary coil of M_2.

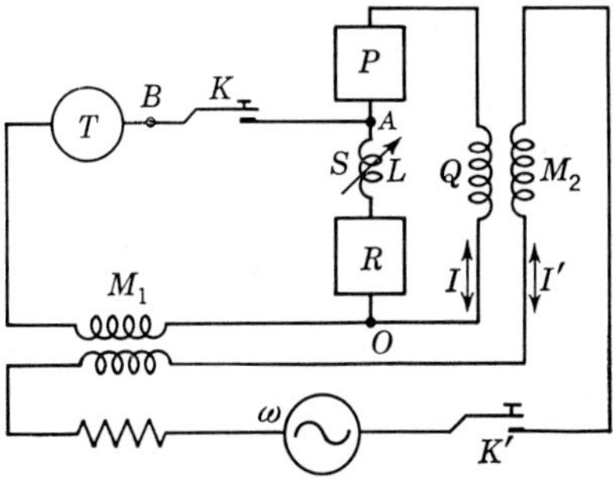

17-12. The Divided-potential Method Using Alternating Current. When alternating current is used in place of the battery shown in Fig. 17-10 it is found impossible to obtain a balance. This is because the current in the circuit RPQ is out of phase with the induced emf, owing to the self inductance in the coil Q. By adding a self inductance L to

FIG. 17-11. Comparison of unequal mutual inductances by the divided-potential method using alternating current.

R the emf over this part of the circuit can be advanced in phase. It is then possible to obtain a good balance against the voltage induced in the secondary coil of M_1, using the arrangement shown in Fig. 17-11.

17-13. Vector Diagrams for Comparing M_1 with M_2 by the Divided-potential Method. The vector diagram for the circuit RPQ (Fig. 17-11) is shown in Fig. 17-12a. The components of the emf required to maintain the current I are laid off in order, giving $ODAJH$. The total voltage OH is $M_2\omega I'$ induced in the secondary coil of the larger mutual inductance M_2.

The induced voltage $M_1\omega I'$ in the other coil is in phase with $M_2\omega I'$ since there is the same current I' in the primary coils. Therefore it is shown as OB (Fig. 17-12a). In the unbalanced case this is not quite the

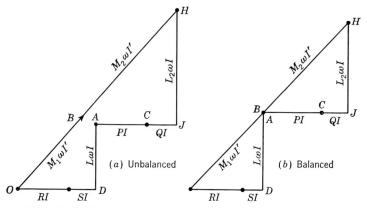

FIG. 17-12. Vector diagrams for the divided-potential method.

same as OA in the main circuit, the difference, BA, appearing across the key K when this is open. When K is closed, there will be current in the telephone T.

By changing L and R, the telephone can be brought to silence. This means that A has been moved to coincide with B, as is shown in Fig. 17-12b for the balanced case. From similar triangles in the balanced condition we have

$$\frac{M_1}{M_2} = \frac{R+S}{R+S+P+Q}$$

17-14. Comparison of a Mutual Inductance with a Capacitance, Using Alternating Current. One of the most satisfactory methods for comparing a mutual inductance with a standard condenser is a modification of the Carey Foster method.

In this method (Fig. 17-13) the induced emf in the secondary of a pair of coils is made to supply the varying charge in the condenser. The main current divides at A', the part I_2 passing through R, and the other part,

I_1, flowing through the condenser C, the resistances N and S, and the coil of self inductance L. The resistance of the secondary coil is included in the value of S, and N includes whatever equivalent resistance there may be in the condenser. At F' the currents unite and pass through P, the primary of the mutual inductance M.

The conditions in this circuit are shown by drawing the diagram for the various emfs, taking these in order along the path $A'B'SF'A'$. Starting with N, and laying off the voltage NI_1 as shown in Fig. 17-14, the voltage over the condenser would be drawn as HB. The voltage over S, which includes the fall of potential in the secondary coil, is drawn parallel to NI_1, since these are in phase, and $L\omega I_1$, is 90° ahead, as shown by QD. There is also in this secondary coil the emf induced by the primary current. This emf is in two parts, which are not in phase with each other. The part of

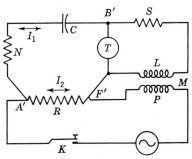

FIG. 17-13. Comparison of a capacitance with a mutual inductance using alternating current.

the current, I_1, that came through S and L continues on through the primary coil and induces in the secondary an emf of the amount

$$e = M \frac{di_1}{dt} = M\omega I_1' \cos \omega t$$

where I_1' denotes the maximum value of the current I_1.

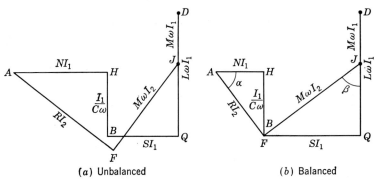

(a) Unbalanced   (b) Balanced

FIG. 17-14. Emf diagrams for the comparison of C and M.

The value of this emf is, then, $M\omega I_1$, and since it is a cosine term, it is either in the same phase as $L\omega I_1$ or directly opposite to $L\omega I_1$, according to the direction that I_1 flows through the primary coil.

The other component of the current, I_2, also flows through the primary coil and gives rise to an emf of $M\omega I_2$ in the secondary, which differs by 90° from RI_2. Since I_2 is behind I_1 by the angle α, $M\omega I_2$ is behind $M\omega I_1$ by the same angle.

Inasmuch as it is desired to obtain a balance, with no current through the telephone circuit, it will be necessary to connect the primary coil into the circuit in the direction that gives $M\omega I_1$ opposed to $L\omega I_1$, and $M\omega I_2$ an angle α behind, as shown in the figure by DJF. The current through R will be behind the condenser current I_1, and therefore RI_2 is represented by AF. As shown above, AFJ is a right angle. For zero emf across the telephone, it is seen that F must coincide with B. As N is decreased, the part $HBQJ$ increases while the right-angled lines JFA connect across from J to A, and F approaches B. As S is varied, B is moved along BQ and can be brought nearer to F. By adjustment of first one and then the other, B can be made to coincide with F, and the telephone shows this balance by silence.

When a balance is obtained, there are two similar triangles in the diagram, and from these,

$$\frac{M\omega I_2}{RI_2} = \frac{SI_1}{I_1/C\omega} = SC\omega = \frac{M\omega}{R}$$

or
$$M = SRC$$

The value of N that is necessary for this balance can be determined as follows: From the diagram,

$$\frac{NI_1}{RI_2} = \frac{L\omega I_1 - M\omega I_1}{M\omega I_2}$$

and
$$N = R\frac{L - M}{M}$$

If the rest of the circuit is well insulated, it may improve the balance to have the point F' grounded.

FREQUENCY

17-15. Measurement of Frequency. A simple and direct method for measuring the frequency of an alternating current is shown in Fig. 17-15a. The primary of a variable mutual inductance M is joined in series with a condenser C and connected to the source whose frequency is desired.

The secondary coil of the mutual inductance is joined in series with a telephone or other detector and connected to the terminals of the condenser. The emf across the condenser is $I/C\omega$, and when properly connected is in opposition to $M\omega I$ in the secondary coil. When M is adjusted to give *minimum* sound in the telephone,

$$M\omega I = \frac{I}{C\omega} \quad \text{and} \quad \omega = \frac{1}{\sqrt{MC}}$$

Usually a balance can be obtained as sharply as the value of M can be read on the scale of the mutual inductance.

If there is a residual hum in the telephone, it is probably due to the power loss in the condenser between the two points where the telephone

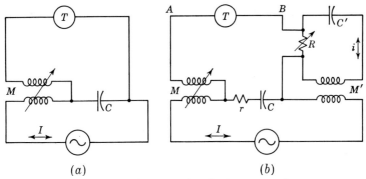

(a)   (b)

FIG. 17-15. Arrangement for measuring the frequency of an a-c source.

circuit is connected. This introduces an unbalanced emf rI into the telephone circuit. If a sharper balance is desired, another emf Ri can be introduced into the telephone circuit, as shown in Fig. 17-15b. C' is a small capacitance and R is a dial resistance that can be varied to make Ri equal to rI.

The vector diagram of the emfs in the telephone circuit when not quite balanced is shown in Fig. 17-16. For a balance A and B are brought together by varying M and R.

17-16. A Frequency Bridge. In most of the arrangements for measuring inductance and capacitance, the balance does not depend upon the frequency of the alternating current except in so far as the quantity being measured varies with the frequency. There are a number of bridge methods, however, in which the frequency is one of the quantities contributing to the balance. When the other factors are known, these methods may be used for the measurement of the frequency.

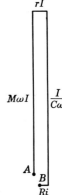

FIG. 17-16. Vector diagram for Fig. 17-15 when not quite balanced.

Thus in Maxwell's method an inductance in one arm is balanced by a capacitance in the opposite arm of the bridge. The balance could have been effected with the condenser in the same arm as the coil, as shown in Fig. 17-17, and with the other arms of noninductive resistances. This means that the inductance and the capacitance must be adjusted to make the resultant effect of the first arm noninductive also. The diagrams for the unbalanced and the balanced conditions are shown in Fig. 17-18, and are self-explanatory.

For the balance

$$L\omega I_1 = \frac{I_1}{C\omega}$$

from which

$$\omega = \frac{1}{\sqrt{LC}}$$

and the bridge is not balanced for other frequencies.   Therefore, with a complex wave form, the harmonics will still be heard in the telephone.

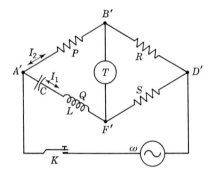

FIG. 17-17. Frequency bridge.

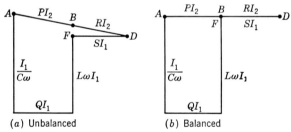

(a) Unbalanced   (b) Balanced

FIG. 17-18. Emf diagrams for the frequency bridge.

Since the frequency f is given by the relation

$$\omega = 2\pi f$$

its value is

$$f = \frac{1}{2\pi \sqrt{LC}}$$

A frequency bridge can thus be made with fixed resistances and a variable self inductance.   In addition to the regular scale, the inductance can be calibrated to read directly the value of the frequency.   By using a few fixed values of C, the use of the bridge can be extended to a wide range of frequencies.

PROBLEMS

17-1. Find the conditions of balance for the bridge shown in Fig. 17-1 using the vector method.

17-2. For the accompanying circuit, show that the bridge can be balanced if $f = \frac{1}{2}\pi R_2 C$.

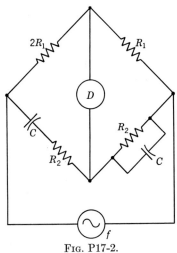

FIG. P17-2.

17-3. Find the conditions of balance for the bridge indicated in the following diagram (Fig. P17-3).

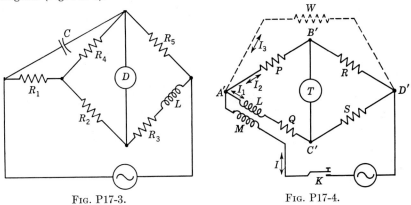

FIG. P17-3.   FIG. P17-4.

17-4. For the balanced mutual inductance bridge shown in Fig. P17-4, prove that $P/Q = R/S$ and $M = L(R/R + S)$.

17-5. Using complex notation, derive the conditions of balance for the frequency bridge shown in Fig. 17-17.

CHAPTER 18

ELECTRONICS

18-1. Introduction. Originally, the term electronics was almost exclusively understood to mean that field of electricity dealing with electron tubes and their associated circuits. The term has now been broadened to include other devices such as crystal diodes and transistors. In fact, these elements have already developed to such a stage that one often encounters circuits which contain no conventional electron tubes at all. One might expect that with the rapid developments in the field of solid-state physics these solid electronic elements will supersede the electron tube in most commercial applications.

18-2. Electron Tubes (Vacuum). The properties of ordinary vacuum tubes arise from the action of electric fields on electrons in free space. It is, therefore, essential to consider the various methods for freeing electrons from matter. It has been pointed out previously that metals conduct electricity as a result of the motion of the so-called "free" electrons. These are electrons in the metal which are bound only very loosely to the atom or to the crystalline structure and may, in fact, be shared or exchanged between atoms. When an electric field is set up within the metal, these "free" electrons drift in a direction opposite that of the applied field and so produce an electric current. Although such electrons can wander about within the metal, they must have a certain minimum energy in order to overcome the retarding potential barrier at the surface of the metal before they can escape into space as *truly* free electrons. This minimum energy of escape is called the work function (ω) of the metal and is commonly expressed in electron volts

$$1 \text{ ev} = 1.6 \times 10^{-12} \text{ erg}$$

The various types of electron emissions can be classified in accordance with the method by which electrons are furnished with this minimum escape energy.

18-3. Thermal Emission. If it were possible to observe the kinetic energy of each of the conduction electrons at a given time, it would be found that, for a given temperature of the metal, distributions would exist as indicated in Fig. 18-1. At room temperature few electrons will have kinetic energies which exceed the work function. However, as the temperature is increased, the distribution shifts to higher energies and it becomes possible for more electrons to escape.

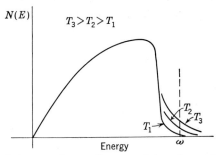

Fig. 18-1. Number of electrons in a unit energy interval as a function of energy at various temperatures. $T_3 > T_2 > T_1$.

The quantitative relationship for thermal emission was first developed by Richardson[1] and can be written

$$i = A T^2 \epsilon^{-\omega/kT} \tag{18-1}$$

where i = emission current expressed, amp/sq cm
A = 120 amp/sq cm-°K² (theoretically constant for all metals)
T = absolute temperature
ω = work function, ergs
k = Boltzmann's constant (1.37×10^{-16} erg/°K)
Typical values of the work function are given in Table 18-1.

TABLE 18-1. WORK FUNCTION FOR SEVERAL TYPICAL METALS

Emitter	ω, ev
Wolfram (tungsten)	4.52
Tantalum	4.3
Molybdenum	4.3
Copper	4.0
Thorium	3.4
Nickel	2.8
Thoriated tungsten	~2.6
Oxide coated	~1

18-4. Photo Emission. When a quantum of light falls on a material and is absorbed, its energy $h\nu$ is given up to the surface. (h is Planck's constant 6.63×10^{-27} erg-sec, and ν is the frequency of the radiation.) Occasionally, a conduction electron near the surface is given this full energy, and if $h\nu$ exceeds the work function for this surface, the electron may escape. Any excess energy goes into kinetic energy of the electron

[1] This relationship is sometimes called the Richardson-Dushman equation.

which has been set free. The number of electrons emitted per second is proportional to the light intensity at a given frequency of radiation.

18-5. Secondary Emission. Energy may also be transferred to an electron within a metal if the surface is bombarded with an external beam of high-energy electrons. Under the proper conditions, it is possible to free several electrons from the metal for each bombarding electron, and as such, this effect can be used as a current amplifier. In commercial photomultiplier tubes photoelectric emission and secondary emission effects are coordinated to produce a very sensitive light-detecting device. Figure 18-2 is a drawing of the essential parts of such a tube. The inner surface of the tube face is coated with a semitransparent layer having high photo efficiency. When light falls upon this surface, some electrons

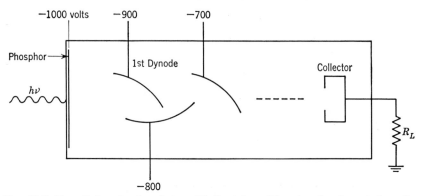

Fig. 18-2. Essential parts of a photomultiplier tube. The phosphor is coated on the inside surface of the glass envelope.

are released. The electrons so released are accelerated by the electric fields between the photo surface and the first multiplying electrode or dynode. Typically, the electrons may be given an energy of the order of 100 volts before striking the first dynode. The dynodes are constructed of materials which are efficient emitters under electron bombardment. This electrode may emit four or five electrons for each one incident upon its surface. Electrons from the first dynode are then directed to the second, where a similar multiplication results. With 10 such stages of multiplication it is possible to realize a total gain of more than 10^6. Two commercial types of photomultiplier tubes are shown in Fig. 18-3.

This tube finds wide application in the field of nuclear physics. It has been found that a great variety of organic and inorganic crystals such as naphthalene, anthracene, stilbene, and sodium iodide (activated with thallium) will emit light when an ionizing particle passes through them. By placing such a scintillating crystal adjacent to the phosphor of a photomultiplier tube, it is thus possible to translate the light pulse into

FIG. 18-3. Photomultiplier tubes. (*Courtesy of Allen B. Du Mont Lab., Inc.*)

a proportionate electrical pulse at the collector of the tube. This provides an extremely efficient detector for gamma radiation as well as for alpha and beta particles. Such scintillation counters have largely replaced the Geiger-Müller counter in nuclear measurements.

18-6. Field Emission. The mechanism of field emission is not completely developed. Empirically, one finds that emission of electrons may be obtained from a cold metal when an electric field of the order of 10^7

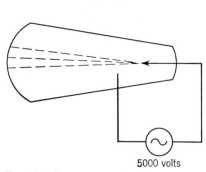

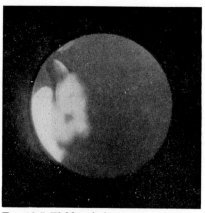

Fig. 18-4. Arrangement for demonstrating field emission. Localized field of 10^7 volts/cm can be set up in the region of the point.

Fig. 18-5. Field emission pattern obtained with arrangement such as Fig. 18-4. The pattern changes continuously as the surface erodes.

to 10^8 volts/cm is applied at the surface of the metal. Apparently the effect is associated with the alteration of the surface potential barrier by the external field. The phenomenon is very striking and can be demonstrated readily with a modified oscilloscope tube. The gun mechanism is replaced by a pointed tungsten wire as indicated in Fig. 18-4. When 5,000 volts is applied between the point and a second electrode, large electric fields are set up in the region of the point and field emission occurs. The electrons thus obtained strike the phosphor, producing a greatly magnified pattern. A photograph of such a pattern is shown in Fig. 18-5.

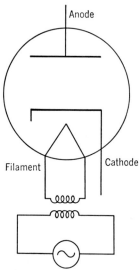

Fig. 18-6. Essential features of a vacuum diode.

18-7. Two-element Electron Tube. The two-element tube or diode can be represented schematically as in Fig. 18-6. Its essential parts are the filament, cathode, and plate or anode. The filament-cathode combination makes up the electron-emitting element. The filament is heated by the passage of a current (alternating or direct) supplied by an external low-voltage supply. The filament could, of course, be used as the emitter of electrons, and this is done in high-power tubes. More commonly, however, the filament is surrounded by a metal tube coated with an oxide emitter which has a much lower work function and, as such, has a higher efficiency. With this arrangement the filament serves simply as a heater for the oxide-coated cathode. Henceforth, the filament will be omitted in drawings of electron tubes.

18-8. Current in a Diode. With a circuit arranged as in Fig. 18-7, let us presume that the heater current is constant and the cathode has reached thermal equilibrium. The cathode is then emitting electrons at a constant rate given by Richardson's equation. One might think that the anode current would rise quickly to the temperature-limited value as given by Richardson's equation and remain at this amount quite independently of the voltage impressed between the cathode and anode. This, however, is not strictly true. As the potential difference is increased, the current rises

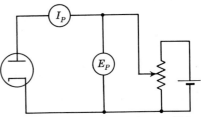

Fig. 18-7. Arrangement for verifying Child's law.

in a manner which can be described by the equation

$$i = \alpha V^{3/2} \tag{18-2}$$

over a wide range of voltages. (This relationship was developed by Child and bears his name. The constant α will depend upon geometric factors.) Eventually, at a high voltage, the current levels off at the temperature-limited value and becomes independent of the impressed potential difference. If the current through the filament is raised, thus raising the cathode temperature, it is found that Child's law again applies even to higher values of the voltage. The leveling off will now occur at a current consistent with the new temperature. This effect, shown in Fig. 18-8, can be understood if one remembers that the electric field between the cathode and anode is influenced by the very presence of the electrons in the space between the electrodes.

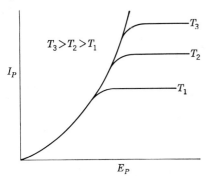

Fig. 18-8. Variation of diode current with voltage. The three-half power law is followed until the space charge becomes depleted.

18-9. Verification of Child's Law. With an arrangement as shown in Fig. 18-7, measure the current-voltage relationship for two diodes with drastically different electrode configurations (6X5 and 6H6). When the data have been plotted, fit the curves with equations of the form $i = \alpha V^x$ to determine the best values of α and x.

Substituting a microammeter to measure the plate current, carefully measure this current when the anode voltage approaches zero. Why does a current still flow when the potential is reduced to zero? Reverse the polarity of the battery and apply a negative potential to the anode until the current vanishes. Would this potential vary with the temperature of the cathode? Investigate this by changing the heating current.

The question now arises as to what will happen when the anode potential is made quite negative (~50 volts). Ideally it should remain at zero. Check this for several tubes both new and old. What factors could contribute to the flow of current with the anode potential negative with respect to the cathode?

18-10. Space Charge. For simplicity, consider the potential distribution which would exist between parallel-plate electrodes without the presence of electrons in the space between them. The electrostatic field would be uniform, and the potential would be a linear function of the distance from the cathode as indicated by line A (Fig. 18-9). If the cathode is now heated so that electrons of low velocity are emitted, the potential distribution will be altered as shown in curve B. For a large emission current the presence of the negative space charge may, indeed, make the potential assume a negative value in the region of the cathode, curve C. In the latter instance the electric field will tend to make the electrons return to the cathode if their initial velocity of ejection will not carry them beyond the minimum of the potential curve. Raising the potential of the anode with respect to the cathode has the effect of reducing the potential minimum close to the cathode and thus makes a greater supply of electrons available to the plate.

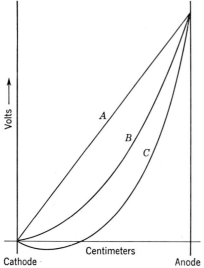

Fig. 18-9. Potential distribution between parallel plate electrodes. Curve A, zero space charge, the potential varies linearly with distance between plates and the electric field is constant throughout the region. Curve B, space charge present, the field in the region of the cathode is reduced. Curve C, large amount of space charge, the field in the region of the cathode is reversed and may drive low-energy electrons back to the cathode.

When the potential is raised sufficiently, the negative of potential will be completely eliminated, in which case every electron emitted will get to the plate. This corresponds to the leveling off of the curves shown in Fig. 18-8.

For electrodes of cylindrical sym-
metry, the corresponding curves are
more complex but the qualitative
description is similar to the case of
parallel plates (Fig. 18-10).

18-11. The Diode as a Rectifier.
At this point it should be obvious
that the diode is a nonlinear element;
that is, Ohm's law does not apply.
Ideally, if the potential is reversed so
that the anode is negative, no current
will flow at all. This property is useful in converting alternating current
into direct current. A simple arrangement for accomplishing this, the
half-wave rectifier, is shown in Fig.
18-11. The diode is connected in

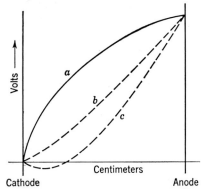

Fig. 18-10. Most of the potential difference between a wire and a concentric
cylinder is near the wire, but a space
charge may greatly modify this distribution.

series with the alternating-voltage source of low impedance and a
resistance R. The output voltage occurring across R is iR. During
the half cycle when the anode is positive, current will flow and a voltage
will be produced across the resistance. During the next half cycle the
anode will be negative, the current will stop, and no voltage occurs across
the resistance. Thus direct, pulsating voltage and current have been produced from an alternating-voltage source.

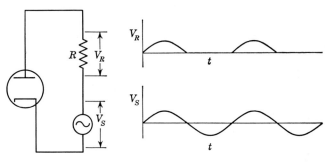

Fig. 18-11. Half-wave rectifier. Electron current flows only during the half cycle
when the anode is positive.

If the internal resistance of the tube is small compared with the
external resistance R, the wave shape will be faithfully reproduced during
the half cycle in which current flows. Conversely, if the internal resistance is not negligible compared with R, the wave form will be distorted.
Kirchhoff's law can be written

$$V_{\text{source}} = V_{\text{tube}} + iR \qquad (18\text{-}3)$$

Thus, if V_{tube} can be neglected, the output voltage reproduces the source voltage. However, if V_{tube} as given by Child's law is appreciable, distortion will result to a certain degree as shown in Fig. 18-12.

In order to make the d-c voltage more constant, it is necessary to add additional circuit elements. One simple procedure is that of placing a condenser in parallel with the load resistance R. Qualitatively this can be looked upon as a method for storing charge and energy during the half cycle when the tube is conducting and feeding the current through the resistor during the nonconducting half cycle. It should be expected that the condenser will be effective for this purpose if the time constant

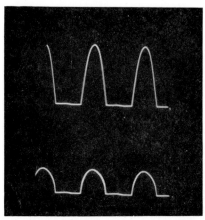
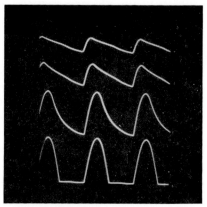

Fig. 18-12. Variation of output wave form with load resistance. Upper curve obtained with high load resistance with little distortion during the conducting half cycle. Lower curve, same arrangement except for much lower value of load resistance. Distortion is evident.

Fig. 18-13. Rectified wave forms obtained by varying the capacitance which is placed in parallel with the load resistance. The time constants RC are 0, 4×10^{-3}, 2×10^{-2}, 4×10^{-2} sec, reading from bottom to top.

RC of this output circuit is great compared with the period of the alternating current. That is, for 60-cycle voltage the product $RC > \frac{1}{60}$. This effect is shown for various time constants in Fig. 18-13.

18-12. Full-wave Rectifier. Diodes used in a full-wave rectifier circuit are shown in Fig. 18-14. This circuit rectifies and takes advantage of both half cycles of the alternating voltage. The source voltage is impressed across a potential dividing resistor or across a transformer which has a center-tap winding. The operation can be seen from the diagram. During the half cycle when point A is positive, electrons will flow from B through the resistance R and the upper tube. During this half cycle the anode of the lower tube is negative with respect to the cathode and is thus inoperative. When the polarity of the source reverses, the tubes exchange roles, with the electron current still flowing in the same direc-

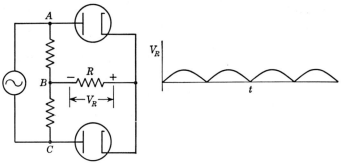

FIG. 18-14. Ordinary full-wave rectifier.   Commonly both diodes are housed in one envelope and share a common cathode.

tion through R.   Clearly this type of wave form can be smoothed more easily than the half-wave type.

18-13. Voltage Regulator Tubes.   Even after the alternating voltage has been rectified and smoothed by an appropriate filter system, one cannot be assured that the magnitude of the direct voltage will remain constant with time.   Without further regulation the direct voltage will change with variations of the alternating line voltage and also with the amount of d-c load drawn from the system.   For small currents a two-element gas-discharge tube provides a simple and efficient method for regulating the voltage.   In structure, the voltage regulator (VR) tube consists of a cylinder with an axial wire.   The tube is filled with an inert gas at a relatively low pressure.   When a voltage is impressed across the tube, current will not flow until a potential is reached at which the gas can be ionized.   After the tube has fired, the current rises rapidly as a function of voltage in excess of the regulating voltage.

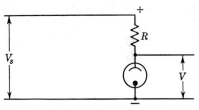

FIG. 18-15. Gas tube arrangement for regulating direct voltages. V_s represents the unregulated direct voltage. V, the voltage across the tube, remains relatively constant with variations of V_s and load current.

Let us assume that above this critical regulating voltage the current varies linearly with the excess voltage:

$$i = \beta(V - V_0) \qquad \begin{cases} V = \text{actual voltage across tube} \\ V_0 = \text{critical regulating voltage} \end{cases} \qquad (18\text{-}4)$$

For a circuit such as shown in Fig. 18-15 we can write

$$V_s = iR + V$$
$$= \beta(V - V_0)R + V$$
$$= \beta V R - \beta V_0 R + V$$

or $$V = \frac{V_s + \beta V_0 R}{1 + \beta R} \qquad (18\text{-}5)$$

Typically $\beta \sim 10^{-2}$ amp/volt

$R \sim 10^4$ ohms

$V_0 \sim 100$ volts

If V_s is small compared with $\beta V_0 R$ and $\beta R \gg 1$, these small terms can be neglected and

$$V \approx V_0 \tag{18-6}$$

independent of V_s.

These tubes are commercially available to regulate at values of V_0 of 75, 90, 105, and 150 volts. The current through the tube should be in the range from 5 to 30 ma. Several types are also available for regulation of voltages up to 900 volts.

With a circuit similar to Fig. 18-15, measure the voltage across the VR tube and the current through it as a function of the source voltage for several values of the dropping resistance R. By what amount could V_s change if the regulated voltage V is to remain constant within 2 per cent?

Place various resistances in parallel with the tube and repeat. (Choose values of the load resistance such that currents of 5, 10, and 20 ma will flow through the load.)

Would you expect this type of regulator to be effective in eliminating high-frequency variations of the voltage?

18-14. The Triode. In considering the influence of space charge it has been seen that the electric field in the region of the cathode plays an important role in determining the amount of current which flows through the tube. Additionally, in the diode a large change in cathode-anode potential is required to change this field distribution appreciably. However, this field can be changed relatively easily by the insertion of a third electrode in the region close to the cathode where the space charge effects are so important (Fig. 18-16). Such an element made of open-mesh wire in order to allow most of the electrons to pass through is called the control

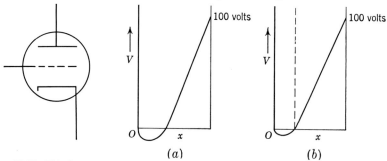

(a)   (b)

Fig. 18-16. Triode vacuum tube. Curve a represents the potential distribution for a diode and shows the influence of space charge near the cathode. Curve b represents the same arrangement except that a metal grid has been inserted and connected electrically to the cathode. This has drastically changed the potential distribution in the region of the cathode.

grid or simply grid. A small change of the grid-cathode potential will then produce a major change in the field shape and greatly influence the flow of electrons to the anode.

18-15. Static Characteristics of a Triode. To describe completely the characteristics of a triode, three variables must be specified. These are the current to the anode or plate I_p, the potential difference between the cathode and grid E_g, and the potential difference between the cathode and anode E_p. A typical set of characteristic curves can be obtained with the circuit shown in Fig. 18-17. The results obtained are similar to those shown in Fig. 18-18. From a consideration of these curves it is now possible to define the three major factors which are of importance in determining the usefulness of a tube for a specific application.

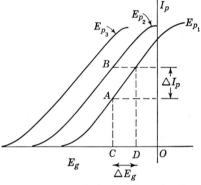

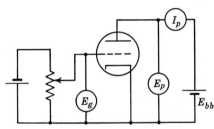

Fig. 18-17. Circuit for measuring the characteristics of a triode.

Fig. 18-18. Typical set of static characteristic curves for a triode.

18-16. Amplification Factor μ. Referring to Fig. 18-18, consider the two independent methods by which the plate current I_p can be changed. First, the grid voltage could be held constant and the plate voltage changed from A to B. This would require a change ΔE_p. Second, the same change in current could be obtained by holding the plate voltage constant and changing the grid voltage from C to D, requiring a change ΔE_g. The amplification factor is defined as the relative effectiveness of the grid and plate in changing the plate current and thus can be written[1]

$$\mu = \frac{\Delta E_p}{\Delta E_g} \qquad (18\text{-}7)$$

[1] An alternate and equivalent definition can be stated as follows: The amplification factor is the negative of the ratio of the change in plate voltage to the change in grid voltage which is required to maintain the plate current at a constant value. That is,

$$\mu = -\left(\frac{\partial E_p}{\partial E_g}\right)_{I_p}$$

The negative sign is required, since it is clear that in order to hold the plate current constant, the change in grid voltage must be in the opposite direction from the change in plate voltage. The resultant value for the amplification factor is a *positive number*.

For a triode having an amplification factor of 20, it would mean that a 1-volt change in the grid potential will change the plate current by the same amount as a 20-volt change in plate potential. It should be noted that the term "amplification factor " does not mean the same thing as the term "amplification" or "gain."

An approximate value for the amplification factor can be obtained from an inspection of the characteristic curves. At the cutoff point the effect of the plate potential is just equal to the opposing effect of the grid potential. Thus[1]

$$\mu \approx - \frac{E_p}{E_g(\text{cutoff})}$$

This value, while easily obtained, may differ considerably from the value which the tube has in the region where the characteristic curves are more linear and, as such, should be taken as only an approximate value for the tube.

18-17. Plate Resistance r_p. This quantity is just what the name implies, the effective resistance between the cathode and anode, or

$$r_p = \frac{\Delta E_p}{\Delta I_p} \tag{18-8}$$

18-18. Transconductance g_m. Conductance, in general, has the form I/V or reciprocal of resistance. Transconductance is defined as the ratio of the change in plate current to the change in grid voltage required to bring about the change in current, or

$$g_m = \frac{\Delta I_p}{\Delta E_g} \tag{18-9}$$

Clearly $$\mu = g_m r_p \tag{18-10}$$

18-19. Dynamic Characteristics of Triodes. The static characteristic curves have been useful in allowing one to define and measure the three important triode characteristics. In application, however, conditions are generally more complex. Commonly it is necessary to insert a load resistance R_L between the plate voltage supply E_{bb} and the plate itself. This has the effect of coupling all three variables E_g, I_p, and E_p so that

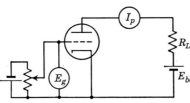

Fig. 18-19. Circuit for measuring the dynamic characteristics of a tube. R_L is the load resistance.

a change in grid voltage will now change both the plate current and plate voltage. In fact,

$$E_p = E_{bb} - I_p R_L \tag{18-11}$$

[1] The negative sign is inserted, since E_g (cutoff) is a negative voltage.

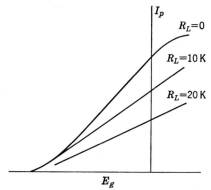

FIG. 18-20. Dynamic characteristic curves. E_{bb} is held constant.

Characteristic curves obtained with the circuit shown in Fig. 18-19 will yield the dynamic properties of the triode. The values of both R_L and E_{bb} are at one's disposal. The main features can be obtained, however, by operating at a fixed value of E_{bb} and using R_L as a variable parameter. Such curves are shown in Fig. 18-20. It will be noted that as R_L is increased, the curves become more linear and have a smaller slope. The curves in Fig. 18-21 show the general variations of μ, r_p, and g_m over a reasonable range of operation.

18-20. Load Lines. The properties of a triode tube can be represented graphically in still another fashion which is sometimes found to be useful. The data which were obtained in Sec. 18-15 can be plotted with I_p and E_p as the variables and E_g as the third parameter (Fig. 18-22). The dynamic characteristics can now be obtained from these static curves by constructing a straight line, the load line, which intersects the abscissa at a plate voltage equal to the supply voltage E_{bb} and which has a

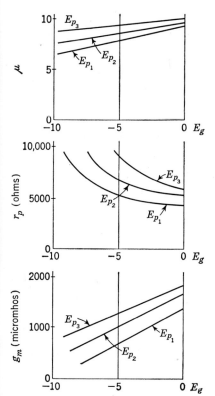

FIG. 18-21. These curves indicate the variations of μ, r_p, and g_m with grid voltage E_g and plate voltage E_p. $E_{p_3} > E_{p_2} > E_{p_1}$.

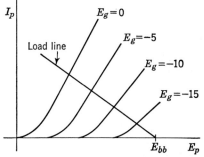

FIG. 18-22. Static characteristic curves for a triode using different representation. This representation also yields the dynamic characteristics.

slope equal to $-1/R_L$. As the grid voltage is changed, the plate current and plate voltage will change but will always lie on the load line so constructed.

18-21. The Triode as an Amplifier. One of the most common uses to which a triode may be put is that of a voltage amplifier. When a small alternating voltage is impressed between the grid and cathode, a relatively large change in plate current takes place. The changing current then produces an alternating component of voltage across the load resistor. Up to this time, when discussing the properties of triodes, we have spoken of direct voltages and currents. These still exist and determine the quiescent operating point of a tube. For example, it may be decided to have a bias of -10 volts on the grid and a voltage E_{bb} of 250 volts for the plate supply. Under these conditions a plate current flows and the

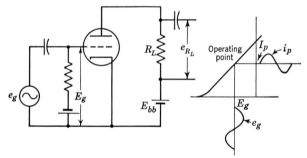

Fig. 18-23. Triode amplifier. The alternating-voltage signal e_g is superimposed on the direct grid bias voltage E_g. The amplified signal is e_{R_L}.

operating point of the tube is determined. Once this is set, these direct voltages are of little further interest. In operating as an amplifier an additional *alternating* voltage is superimposed upon this direct bias voltage. This produces an *alternating* component of plate current superimposed on the direct plate current. The alternating plate current further produces an *alternating*-voltage drop across the load resistor R_L superimposed on the direct-voltage drop. It is these alternating components with which we shall be most concerned in the future. We shall use the symbols e_g to represent only the alternating component of grid voltage, i_p the alternating anode current, and e_p the alternating anode voltage. These values are shown graphically in Fig. 18-23.

18-22. Amplification or Gain of a Triode Amplifier. The gain of an amplifier is defined as the ratio of the output voltage to the input voltage, or e_{R_L}/e_g for the circuit shown in Fig. 18-23. We shall further assume that the tube is operating on a linear portion of its characteristic curve and μ, r_p, and g_m can be considered constant in the range of operation.

When a signal e_g is applied to the grid, the current through the tube will change owing to two effects: (1) e_g by itself will cause a change in current; (2) the change in current through the load resistor will cause the plate voltage to change in a direction opposite that of the grid. Since the grid is more effective than the plate by a factor μ, we can combine both these effects and express them in terms of an equivalent plate-voltage change. Applying Ohm's law to the tube we have

$$\mu e_g - e_p = i_p r_p \qquad (18\text{-}12)$$

where μe_g = effect of grid-voltage change in terms of equivalent plate-
voltage change
e_p = actual plate-voltage change
but in our circuit $e_p = i_p R_L$ (since the tube resistance is assumed to be constant). Thus

$$i_p = \frac{\mu e_g}{r_p + R_L} \qquad (18\text{-}13)$$

Multiplying by R_L we have

$$i_p R_L = \frac{\mu e_g R_L}{r_p + R_L}$$

The gain therefore is $i_p R_L / e_g$, which equals

$$\frac{\mu R_L}{r_p + R_L} \qquad (18\text{-}14)$$

From this equation the relationship between gain and amplification factor becomes clear. The maximum gain attainable from the tube is μ, and this occurs when $R_L \gg r_p$.

18-23. Equivalent Circuits for the Triode. From an inspection of Eq. (18-13) it is apparent that the tube acts as a generator having a voltage μe_g and a resistance r_p as shown in Fig. 18-24. This is called the constant-voltage equivalent circuit.

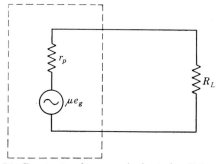

FIG. 18-24. Constant-voltage equivalent circuit for a triode.

Again referring to the output voltage $= \mu e_g R_L/(r_p + R_L)$ and substituting $g_m r_p$ for μ, we have

$$\text{Output voltage} = g_m e_g \frac{r_p R_L}{r_p + R_L} \tag{18-15}$$

From this we see that the tube can be considered as a generator feeding a

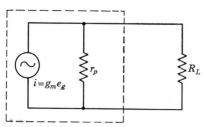

FIG. 18-25. Constant-current equivalent circuit for a triode.

current $g_m e_g$ into a parallel combination consisting of the plate resistance in parallel with the load as in Fig. 18-25. This is the constant-current equivalent of a tube.

18-24. Frequency Response of a Real Amplifier. In the elementary discussion of the amplifier, certain complications were neglected for the sake of clarity. A more realistic circuit is shown in Fig. 18-26. In order to utilize the amplified alternating voltage, an isolating or coupling condenser C_c is ordinarily used to transmit the alternating signal while blocking the direct plate voltage. r_g represents the resistance across which the alternating voltage is developed. It may, for example, represent the grid resistance of a second stage of amplification. C_p represents a capacitance shunting the load resistance arising from plate to cathode capacitance in the tube as well as stray wiring capacitance. Similarly

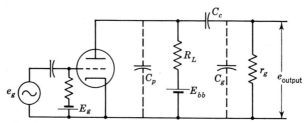

FIG. 18-26. Circuit representing a realistic amplifier. C_p is the stray capacitance from plate to cathode. C_g represents the stray capacitance across the output resistance.

C_g represents a capacitance shunting the output resistance. This again may arise from wiring and grid-cathode capacitance of the second stage of amplification.

The complete equivalent circuit for such an amplifier is shown in Fig. 18-27. Rather than work out a single solution for the gain of this circuit, it is more useful to consider three separate important cases.

Before investigating the numerical values of the gain, it is necessary to consider the magnitudes of the various elements involved. Typically

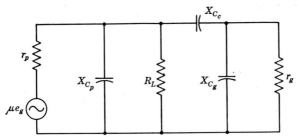

FIG. 18-27. Equivalent circuit for a real amplifier.

R_L may be of the order of 10^4 ohms, $r_g \sim 10^5$ ohms, $C_p \sim 10^{-11}$ farad, $C_g \sim 10^{-11}$ farad, $C_c \sim 10^{-6}$ farad.

18-25. Mid-frequency Range. For the values given above, if we choose frequencies to be amplified in the range from 100 to 10^5 cycles per sec, it is seen that X_{C_c}, which is in series with the output resistance, varies in the range between $\sim 1,000$ and 10 ohms. Since r_g is $\sim 10^5$ ohms, X_{C_c} can be neglected as far as the alternating component is concerned. X_{C_p} and X_{C_g} are in parallel with R_L. Again in this frequency range X_{C_p} varies between 10^8 and 10^5 ohms. Since this high impedance is in parallel with $R_L \sim 10^4$ ohms, both X_{C_p} and X_{C_g} can be neglected. The equivalent mid-frequency circuit is seen in Fig. 18-28. For this circuit

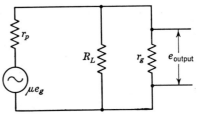

FIG. 18-28. Equivalent circuit for an amplifier in the mid-frequency range.

$$i = \frac{\mu e_g}{r_p + r_g R_L / (r_g + R_L)}$$

and
$$e_{\text{output}} = i \frac{r_g R_L}{r_g + R_L} = \frac{\mu e_g r_g R_L}{r_p r_g + r_p R_L + r_g R_L} \qquad (18\text{-}16)$$

Substituting $\mu = g_m r_p$,

$$e_{\text{output}} = \frac{g_m e_g r_p r_g R_L}{r_p r_g + r_p R_L + r_g R_L}$$

The gain is

$$\frac{e_{\text{output}}}{e_g} = g_m R_{\text{parallel}} \qquad (18\text{-}17)$$

where R_{parallel} is the equivalent resistance of r_p, r_g, and R_L all in parallel.

Clearly the gain in this mid-frequency range is constant within the region of validity of the assumption that the capacitive reactances can be neglected.

18-26. High-frequency Range. At higher frequencies the assumption concerning the reactance of the coupling capacitor becomes better. It is no longer legitimate to neglect the shunt capacitors, however. The equivalent circuit is shown in Fig. 18-29. The solution will be carried out using the constant-voltage equivalent circuit. If the total shunt reactance X_{C_s} is broken off at A-B, Thévenin's theorem can be applied.

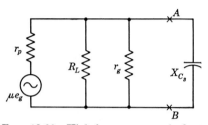

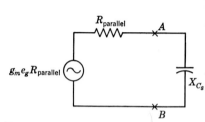

FIG. 18-29. High-frequency equivalent circuit. Series capacitances can be neglected, but shunt capacitances are important.

FIG. 18-30. Thévenin's equivalent for high-frequency approximation.

The circuit to the left is replaced in the usual manner (Fig. 18-30). The current is

$$i = \frac{g_m e_g R_{\text{parallel}}}{(R_{\text{parallel}}^2 + X_{C_s}^2)^{1/2}} \tag{18-18}$$

and

$$e_{\text{output}} = \frac{g_m e_g R_{\text{parallel}} X_{C_s}}{(R_{\text{parallel}}^2 + X_{C_s}^2)^{1/2}}$$

$$\text{Gain} = \frac{g_m R_{\text{parallel}}}{[1 + (R_{\text{parallel}}/X_{C_s})^2]^{1/2}} = \frac{\text{mid-frequency gain}}{[1 + (R_{\text{parallel}}/X_{C_s})^2]^{1/2}} \tag{18-19}$$

Thus it is seen that the shunting effect of the capacitors has the influence of reducing the gain at high frequencies.

At a frequency such that R_{parallel} equals X_{C_s}, the gain will have been reduced by the factor $1/\sqrt{2}$. This is called the upper half-power frequency.

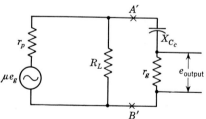

FIG. 18-31. Equivalent amplifier circuit for low frequencies.

18-27. Low-frequency Range. As the frequency is decreased below the mid-frequency range, the shunting effect of the capacitors is again negligible. It is important, however, to consider the influence of the series coupling capacitance which now will combine with the output resistance r_g to form a potential divider or filter. The equivalent circuit is given in Fig. 18-31. Again Thévenin's theorem is applied (Fig. 18-32). The output voltage is

$$ir_g = \frac{g_m e_g r_p r_g R_L}{(r_p + R_L)\{X_{C_c}^2 + [r_g + r_p R_L/(r_p + R_L)]^2\}^{1/2}} \tag{18-20}$$

Rearranging,

$$e_{\text{output}} = \frac{g_m e_g r_p r_g R_L}{(r_p r_g + r_g R_L + r_p R_L)\left\{1 + \left[\dfrac{X_{C_c}}{r_g + r_p R_L/(r_p + R_L)}\right]^2\right\}^{\frac{1}{2}}}$$

$$\text{Gain} = \frac{\text{mid-frequency gain}}{\left\{1 + \left[\dfrac{X_{C_c}}{r_g + r_p R_L/(r_p + R_L)}\right]^2\right\}^{\frac{1}{2}}} \qquad (18\text{-}21)$$

Again the amplification is reduced, this time owing to the series impedance of the coupling condenser.

A typical gain versus frequency curve is shown in Fig. 18-33.

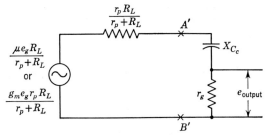

FIG. 18-32. Thévenin's equivalent for a low-frequency amplifier circuit.

The gain versus frequency for an amplifier as shown in Fig. 18-26 can be conveniently carried out with the usual laboratory equipment if the high-frequency end is reduced to the range of the oscillators which are available for the signal. (This is usually 20,000 cycles.) As mentioned previously, the upper half-power point occurs when R_{parallel} equals X_{C_s}. When the shunt capacitance is artificially increased by placing an external capacitance from anode to cathode, the upper part of the gain curve can be brought into the range of the instruments which are available.

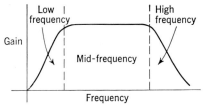

FIG. 18-33. Typical gain versus frequency relationship for an amplifier of the type discussed.

18-28. Grid Bias for Triode Amplifier. In order for the output voltage to reproduce the input signal in wave form, it is necessary to operate the grid voltage somewhat above the cutoff potential of the tube and never allow it to be driven positive. The lower limit is imposed by the non-linearity of the characteristic curves in the region of cutoff, while the upper limit is brought about by the fact that grid current will flow if the grid becomes positive. While often a small amount of grid current can be tolerated, it often reduces appreciably the effective grid signal, especially if the source of the signal has a high impedance. An example of these effects is shown in Fig. 18-34.

Up to this time we have considered that the quiescent operating point of the tube was produced by a bias battery in the grid circuit. This arrangement is not convenient. It is equally possible to operate the grid at ground potential and raise the cathode above this to set the grid at a negative voltage with respect to the cathode. This can be done conveniently by inserting a cathode resistance R_k between the cathode and the negative side of the plate supply. All the plate current then flows through this resistance and sets the cathode at a positive potential.

$$E_k = I_p R_k \tag{18-22}$$

The resistance by itself is not enough, however, since a signal at the grid causes a variation in plate current. This would also change the grid bias and reduce the effective grid signal. This defect is overcome by placing a "large" condenser C_k in parallel with R_k. (By "large" we mean a condenser whose reactance

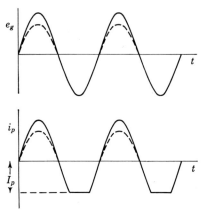

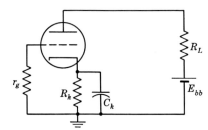

FIG. 18-34. Distortion of wave forms. Upper, when the grid potential becomes positive with respect to the cathode, electrons will flow to the grid and lower the effective grid voltage (dotted portion). Lower curve shows alternating plate current. A large negative signal causes a complete cutoff of current as indicated.

FIG. 18-35. Self-biasing arrangement for amplifier. The direct current flows through R_k to produce a direct grid bias voltage. The alternating current is bypassed through the low impedance of C_k and thus does not influence the direct bias.

X_{C_k} is ~10 per cent of R_k at the lowest frequency for which the amplifier will be used.) This condenser will have a low impedance for the alternating component and an infinite impedance for the quiescent or direct component. Thus the bias will remain constant and equal to $I_p R_k$. Such a circuit is shown in Fig. 18-35.

18-29. Influence of the Physical Structure of the Triode on μ. As mentioned previously, the amplification factor is a measure of the relative effectiveness of the grid voltage and plate voltage in producing a change in plate current.   Clearly the position of the grid with respect to the cathode will influence its effectiveness appreciably.   The plate is effective in changing the current only in so far as the electric field produced by the plate "leaks" through the grid structure into the region of high space charge.   Obviously the physical construction of the grid can be of importance in determining this leakage field.   For example, an open grid structure with wide spacing of wires will allow the plate to have a large influence in the space-charge region and thus will lead to a low amplification factor.   On the other hand, a grid with closely spaced wires will more effectively shield this important region and produce a higher amplification factor.

18-30.   The Pentode Tube.   A more effective method for shielding the cathode from the plate is to insert one or more additional grids between

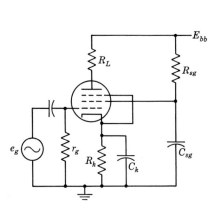

Fig. 18-36. Typical pentode amplifier.

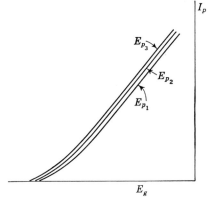

Fig. 18-37.   Static characteristics of a pentode. $E_{p_3} > E_{p_2} > E_{p_1}$.

the control grid and the anode.   A pentode tube employs two additional grids held at constant potentials.   The grid next to the control grid, called the screen grid, is usually operated at a fixed positive potential of the order of 60 volts.   An additional grid is placed between the screen grid and the anode.   This is the suppressor grid.   Its potential is normally that of the cathode.   The purpose of the suppressor is chiefly that of eliminating harmful secondary emission effects.   A typical amplifier using a pentode is shown in Fig. 18-36.   A set of characteristic curves is given in Fig. 18-37.   A comparison of the properties of a triode and pentode is made in Table 18-2.

The potential for the screen grid is normally obtained from the same power supply as the plate voltage simply by inserting a resistance of sufficient value that the screen grid current will produce enough voltage drop in this screen grid resistance.   Here again it is essential to insert a bypass condenser from the screen grid to ground so that the potential of this grid will not be affected by the alternating signal.

TABLE 18-2. COMPARISON OF TRIODE AND PENTODE CHARACTERISTICS

	Triode	Pentode
μ	~10	~500
r_p	~10^4 ohms	~5×10^5 ohms
g_m	~10^3 micromhos	~10^3 micromhos

18-31. Equivalent Pentode Circuits. For a pentode amplifier, with properly bypassed cathode and screen grid, the equivalent circuits and equations for the gains are identical with those already computed for the triode. However if bypassing of one or both of these elements is not adequate, the low-frequency gain will, in general, drop off more rapidly as the frequency is decreased.

18-32. Vacuum-tube Voltmeter. Generally an a-c voltmeter of the conventional portable type has a low resistance. This can, of course, lead

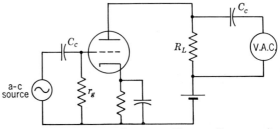

FIG. 18-38. Circuit for vacuum-tube voltmeter. The coupling capacitances C_c should have a low impedance at the frequencies for which the meter is to be used.

to difficulties in measuring voltages from a high-impedance source. It is possible to overcome this by using an electron tube as an amplifier and impedance transformer as indicated in Fig. 18-38. The grid resistance, which is the load resistance for the alternating source, can be made many megohms, so that the current supplied by the source may be ~10^{-8} amp or less. The current in the load resistance typically may be many milliamperes. This circuit can be calibrated by means of a low-impedance source.

18-33. The Oscilloscope. One of the most useful electronic measuring devices is the cathode-ray oscilloscope (Fig. 18-39). Its three main parts are the electron gun, the horizontal deflection plates, and the vertical deflection plates. The electron gun projects a narrow beam of high-speed electrons toward a fluorescent screen, a visible spot being produced where the beam impinges upon the screen. In traversing the path

FIG. 18-39. Cathode-ray tube gun and deflection plates. Oscilloscope. (*Courtesy of Tektronix, Inc.*)

between the electron gun and the fluorescent screen, the beam passes through two sets of parallel deflection plates arranged at right angles to each other. The electric field between one set of plates produces a horizontal deflection of the electrons, while the second produces a vertical deflection. Normally a saw-tooth voltage is applied to the horizontal deflection plates, causing the beam to move horizontally at a constant rate. When the beam has been deflected the full width of the tube, the deflecting voltage drops suddenly to return the beam to its starting point and the process is repeated. An external voltage to be studied can be applied to the vertical deflection plates, usually after being amplified to produce a larger vertical deflection. The wave form of the external source is thus visually displayed on the face of the oscilloscope.

18-34. The Synchroscope. Basically, the synchroscope is similar to the ordinary oscilloscope in construction and operation. The two instruments differ in their horizontal deflection systems. The deflection in the oscilloscope is caused by an oscillator which produces a saw-tooth wave at regularly recurring intervals. To produce a steady pattern, the signal to be studied must recur at a regular frequency and be synchronized with the horizontal sweep. The synchroscope deflecting system, on the other hand, is triggered by the incoming signal itself, so that randomly recurring signals can produce a steady pattern.

18-35. Tuned Amplifier. We have already discussed the operation of amplifiers in which the load is resistive. These could be classified as

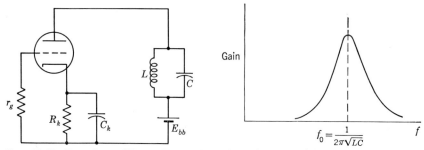

Fig. 18-40. Tuned amplifier. Gain will be large in the neighborhood of resonant frequency.

linear or wide-band amplifiers. If the load is changed from a resistance to a parallel resonance circuit, as in Fig. 18-40, the frequency characteristics are changed drastically. The gain takes essentially the same shape as the impedance of the LC load circuit. This type of circuit is particularly useful when it is desirable to amplify only a narrow band of frequencies while rejecting higher and lower values.

18-36. Oscillator. Since an electron tube can amplify both voltage and power, it is possible to set up a system in which a small part of the output voltage is fed back to the input, in the proper phase, so as to sustain the oscillations. The alternating power thus obtained is, of course, supplied from the direct-voltage plate supply. The load is again an antiresonance circuit which determines the frequency of oscillation to a large extent. The feedback from plate to grid can be accomplished by con-

necting a small capacitor between these
two elements as shown in Fig. 18-41.

18-37. Gas Diode. If a small
amount of inert gas or vapor is ad-
mitted to an electron tube which has
basically a diode structure, it is found
that the properties are changed radi-
cally. This change is brought about
by the fact that atoms of the gas can
be ionized by collisions with the electrons, which are accelerated in their
transit between electrodes. The positive ions, being much heavier than
the electrons, have a much lower

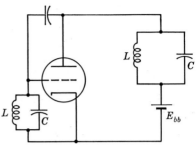

FIG. 18-41. Vacuum-tube oscillator.
The natural capacitance between plate
and grid is often adequate, so that an
external feedback capacitor is not
required.

mobility and remain in the interelectrode space for a long period of time.
Their main effect is to neutralize effectively the negative space charge
and thus permit the current to reach its limiting value at a much lower
voltage. A comparison of the current-voltage relationships is shown in
Fig. 18-42 for both the gas and vacuum diodes. The curves for each tube
are identical at the very lowest voltage. This is the case since at these
voltages the electrons do not have sufficient energy to produce ionization
in the gas and thus space-charge effects are still present in the same
manner in each tube. Once the ionization potential ($\sim$15 volts) is
reached, positive ions are formed, the space charge is neutralized, and the
current rises abruptly to the value limited by the cathode emission.

The neutralizing effect of the positive ions gives the gas diode several
very important advantages over the
vacuum diode. Since space charge
need no longer be considered, it is
possible to use emitters (usually a
filament) having a very large area
made by kinking the heated element.
This would have no advantage in a
vacuum diode, since space-charge
effects would prevent the electrons
from escaping from the inner parts
which are kinked. With the gas
diode, the positive ions enter these
narrow regions and permit the whole
emitter to be effective. It is therefore possible to have large usable
emission currents from a filament
which occupies a small volume.

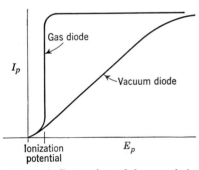

FIG. 18-42. Comparison of characteristic
curves for a vacuum diode and a gas
diode of the same structure. Positive
ions are formed when the electrons
reach an energy equal to the ionization
potential of the gas. Typically this
may be $\sim$15 volts.

The gas diode generally will be more efficient than a corresponding vacuum diode, since the voltage drop across the tube always remains low, even for large currents.

18-38. Gaseous Triode. It has been seen that a small amount of gas when added to a diode tube can drastically change its characteristics. Likewise, when an inert gas or vapor is added to a triode-type tube, its properties are also changed. This can be illustrated by means of the diagram (Fig. 18-43). Assume that the cathode is heated and the grid potential set at point *A* (lower than cutoff) before the plate voltage is applied. After the anode or plate voltage is applied, the grid potential is raised gradually. No plate current will flow until the cutoff potential at *B* is reached. This value is determined by the plate potential and structure of the tube just as in a vacuum triode. Once the cutoff potential is reached, the plate current suddenly jumps to a high value indicated by point *C*. Increasing the grid potential further does not change the plate current. Upon decreasing the grid potential it is found that the plate current remains constant even if the grid potential is lowered far beyond the cutoff potential.

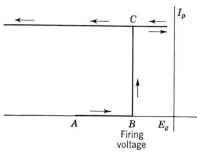

FIG. 18-43. Characteristics of gaseous triode. After the tube has fired, the grid loses its control and the current rises to a value limited by the temperature of the cathode.

The qualitative aspects of this operation can again be understood in terms of the influence of positive ions within the tube. Initially the grid prevents electrons from being accelerated by the high plate potential, and no current will flow. Once the tube starts to conduct (at point *B*), positive ions are formed and are attracted by the negative grid. These heavy positive ions effectively form a sheath about the grid and eliminate the influence of the grid potential completely. From this point on, the tube acts in a manner similar to a gaseous diode. The grid cannot gain control by itself even if its potential is reduced by several hundred volts. The tube can be returned to its initial nonconducting state, however, by lowering the plate potential to a point below the ionization potential of the gas. This quickly eliminates the positive ions, and the grid can then again take control. It is apparent that such a characteristic curve may be of use in triggering devices.

18-39. Gas-tube Relaxation Oscillator. In Fig. 18-44 we have indicated the essential parts of a gas-triode relaxation oscillator. Consider the tube to be initially biased below cutoff with the switch *S* open. When the switch is closed, the plate potential will rise, at all times being the potential of the condenser. Since the condenser C is being charged through the resistance R, the charging time constant will be RC. As the plate potential rises, the cutoff potential of the tube becomes more negative and eventually reaches the applied grid potential E_g. At this point the tube fires and discharges the condenser through the low effective resistance of the tube. Because of the small amount of stray inductance in the connecting wires and tube, this discharge is similar to that discussed in Sec. 13-8. The potential will drop and actually shoot below the ionization potential, whereupon the grid will regain control and the process will be repeated. When a high-voltage supply E_{bb} is used and the grid adjusted to fire at a low value of E_p, the charging curve can be made much

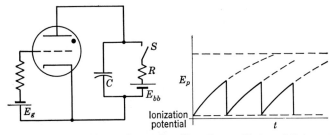

FIG. 18-44. Circuit and wave form of a relaxation oscillator of the gas type.

like a saw-tooth wave, and in fact, this type of oscillator is often useful in producing the horizontal sweep of an oscilloscope. The repetition frequency can be varied by adjusting R or C or both.

NONVACUUM-TUBE ELECTRONIC ELEMENTS

18-40. Conduction in Solids. For purposes of classification it is convenient to consider three types of solids: the insulator, semiconductor, and conductor. A good insulator may have a specific resistance of 10^{15} ohm-cm, while a good conductor may have a value of 10^{-5} ohm-cm, yet the basic structures of the atoms are not greatly dissimilar. In order to explain these great differences it is necessary then to look not only at the properties of the atom but also at the properties of the solid of which the atom is a part.

18-41. Energy Bands. In a solid, we can think of the electrons as existing in discrete energy bands in a manner analogous to the discrete energy levels of the isolated atom. The innermost, tightly bound electrons are in the lowest levels, while the valence electrons which form the bonds with the remaining atoms of the solid fill the uppermost band. All these electrons are bound in the structure of the atom or solid and are not free to move about within the solid. In addition to these bands, which are normally filled, there is a higher empty energy band called the conduction band. If, by thermal agitation or some other means, an electron in the valence band can make the energy transition to the conduction band, it will then be a "free" or conduction, electron.

It is now possible to classify solids with respect to their electrical properties. An insulator gains its properties by having its conduction band many volts above the top of the valence band, and as such there is little probability that an appreciable number of electrons will reach this state at ordinary temperatures. Of course, as the temperature of an insulator is raised, there is a greater chance that more electrons will be promoted to the conduction band and the conductivity of an insulator will thus increase with temperature.

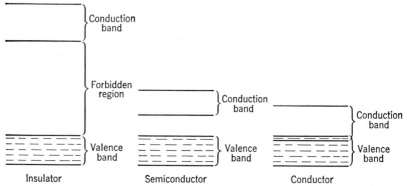

FIG. 18-45. Representation of valence and conduction bands for insulator, semiconductor, and conductor.

A semiconductor differs from an insulator in that its conduction band is only slightly higher than the top of its valence band, so that even at normal temperatures some electrons are promoted to the conduction band. This separation of bands might be of the order of 1 volt or less.

In a conductor, the conduction and valence bands are extremely close and, in fact, commonly overlap, so that one will always find conduction electrons which are free to drift in a fixed direction when an electric field is applied. These bands are illustrated in Fig. 18-45.

18-42. Charge Carriers in an Ideal Semiconductor. When a valence electron in a semiconductor solid is raised to the conduction band, there are two methods by which charge can be transported within the solid. First, the conduction electron itself will drift in the direction opposite the applied electric field and thus transport charge. Second, there will be a vacancy or hole in the valence band from which the electron departed. Referring to Fig. 18-46, it will be noted that the electrons to the left of of the hole will be urged to move to the right to fill the vacancy, and effectively this moves the hole in the direction of the field. For all practical purposes the hole may be looked upon as having a positive charge equal in magnitude to that of the electron and flowing to the left under the influence of the electric field. The total charge transported will thus be made up of that carried to the right by the conduction electron plus that carried to the left by the moving hole. Of course, it is possible to de-excite the conduction electron by collision, in which case it may fall back to the valence band to recombine with a hole. The excess energy may be carried off in the form of radiation.

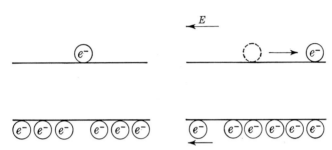

Fig. 18-46. Transport of charge in a semiconductor. The electron in the conduction band is urged to move in the direction opposite the applied field. In addition, the hole or vacancy in the valence band moves in the direction of the field.

18-43. Effect of Impurities in Semiconductors. Due to the relatively high energy gap between the valence and conduction bands of a semiconductor, the conductivity in the pure state is very low at ordinary temperatures. To be specific, for germanium, the energy gap is 0.7 ev while the thermal energy available at room temperature is less than 0.03 volt. This situation may be changed drastically, however, by the introduction of controlled amounts of impurities. For example, the introduction of 1 part in 10^5 of impurities may increase the conductivity by a factor of several hundred. We shall next inquire as to the nature of the physical effects which can produce such drastic changes in electrical properties of solids.

18-44. Donor Impurities. Germanium (which probably has been studied in more detail than any of the other semiconductors) has four valence electrons and crystallizes in a form such that each atom is surrounded by four other atoms. One of the valence electrons from each atom is paired to each other neighbor, thus forming four covalent bonds. If an atom such as antimony, which has five valence electrons, is now substituted in the crystal structure in place of a germanium atom, the four covalent bonds can be formed and the fifth electron will remain unbonded in the structure. This fifth electron will still be bound loosely (~ 0.05 ev) to the impurity atom. It is to be expected, however, that many of these electrons will be found in the conduction band at all times owing to their thermal energy. This effect is indicated in Fig. 18-47. Germanium

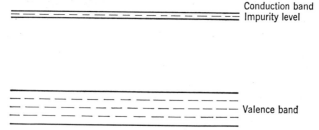

FIG. 18-47. Energy bands for *n*-type semiconductor. The excess electrons from the donor impurities are bound very loosely and may reach the conduction band by thermal agitation even at room temperature.

with this type of impurity will have an excess of negative charge carriers and is called an *n*-type semiconductor.

18-45. Acceptor Impurities. The conductivity of semiconductors can also be increased by the introduction of acceptor impurities such as boron. These impurities have just three valence electrons, and when introduced into a germanium-type crystal structure, they are able to form only three covalent bonds with neighboring atoms. The imperfection thus produced may be thought of as being attached to the positive core of the impurity atom. Since the imperfection consists of the lack of one electron attached to a positive core, one might expect that an energy level slightly higher than the top of the valence band may be assigned to the presence of this imperfection. At normal temperatures some of the covalent electrons in the valence band will be excited to this level, leaving a positive hole in the valence band which then can move under the influence of an electric field as described previously. In this type of crystal the charge carrier is the positive hole, and it is therefore called a *p*-type semiconductor.

18-46. A *p-n* Junction. A semiconductor crystal formed with a donor-type impurity on one side and an acceptor type on the other, while still retaining the continuity of the crystalline structure, produces a *p-n* junction. Electrons from the *n*-type crystal will diffuse into the *p*-type

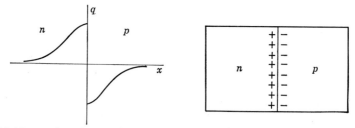

FIG. 18-48. *p-n* junction. Diffusion of holes and electrons across the boundary produces a double layer of charge as indicated.

and become attached, or bonded, into the crystal structure.   (One could also say that the electron combined with and neutralized a positive hole.) This leaves the p-type material with an excess of bound negative charge in the neighborhood of the boundary.   At the same time positive holes will diffuse into the n-type medium, and these will neutralize the conduction electrons in the donor material. This leaves the n-type medium with an excess of bound positive charge in the immediate neighborhood of the junction.   This action continues until the double charge layer thus produced is large enough to prevent further diffusion (Fig. 18-48).

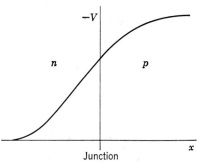

FIG. 18-49. The potential distribution in the region of the p-n junction is caused by the diffusion of electrons and holes.

The presence of bound charges at the junction sets up a potential difference in this region which can be represented as shown in Fig. 18-49. An electron now cannot move from the n to the p region unless it has sufficient energy to carry it over this potential barrier at the junction.

18-47. Rectifying Action at a Junction.   The rectifying action of a crystal junction, such as described, can be understood if the potential

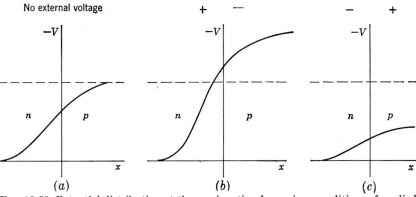

FIG. 18-50. Potential distribution at the p-n junction for various conditions of applied external voltage.   The height of the barrier which the electron (or hole) must climb depends upon this external potential.   The dashed line represents the thermal energy available to the electrons.

change across the junction is investigated under several conditions of applied voltage (Fig. 18-50).   With no external voltage applied to the crystal, the negative potential at the barrier is just sufficient to keep further electrons from entering the p region and holes from entering the n region. When an external voltage is applied such as to make the p region more negative, the barrier height is increased and again the major carriers cannot cross the boundary. Effectively the junction shows a high resistance under these conditions, just as a vacuum diode has a high resistance when its anode is negative with respect to the cathode.

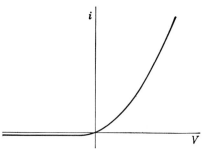

FIG. 18-51. Characteristic curve for a p-n junction diode. The forward resistance is several orders of magnitude less than the reverse resistance.

An external potential connected so as to make the p region more positive than the n will, of course, reduce the barrier which the electrons must overcome, and a large electron current will flow from the n to the p region in the forward direction. In such crystals the holes also migrate in the direction opposite that of the electrons. The description of their action follows that of the electron. The total current is then just the sum of the electron current and the hole current. A typical characteristic curve for such a diode is shown in Fig. 18-51.

18-48. Transistor Actions. In its simplest form a transistor of the junction type consists of a crystal of semiconductor such as germanium which has been doped with impurities to form three alternate n-type and p-type regions as shown in Fig. 18-52. The transistor shown is of the n-p-n type. The p region, although represented as thick, is actually

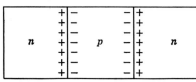

FIG. 18-52. n-p-n transistor showing the charge distribution at the junctions.

FIG. 18-53. Potential distribution in an n-p-n transistor arising from the migration of electrons and holes.

kept extremely thin. The charge distribution at such junctions has already been discussed. The potential difference across the junctions with no outside voltage applied is represented in Fig. 18-53.

Let us call the section to the left the *emitter*, the middle section the *base*, and the right section the *collector*. If an external circuit is connected to the transistor as shown in Fig. 18-54, the potential distribution for the electrons will be correspondingly altered. The emitter-base junction is biased in the forward direction, so that electron current will flow easily between these two elements. Once the electrons get into the base, or p, region, the direction of the potential at the base-collector junction is such as to urge them into the collector. The magnitude of this electron current will be very dependent upon the height of the potential barrier at the emitter-base junction.

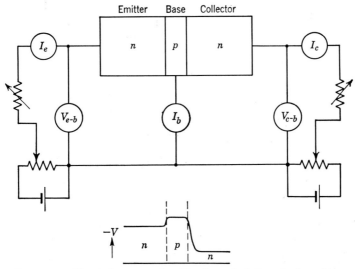

Fig. 18-54. Circuit for measuring the characteristics of a transistor.

18-49. Determination of the Characteristics of a Transistor.
Five quantities can be used in describing the properties of a transistor. These are the emitter-base potential difference V_{e-b}, the collector-base potential difference V_{c-b}, the emitter current I_e, the base current I_b, and the collector current I_c. If any two of these are known, the others are also determined. Commonly, a characteristic curve family is plotted with I_c as the abscissa and V_{e-b} as the ordinate for various values of I_e. This can be carried out with the circuit shown in Fig. 18-54. Load lines can also be plotted for these curves.

18-50. The Junction Transistor as a Voltage Amplifier.
The junction transistor does not amplify current. In fact I_c is always slightly less than I_e. It can, however, be used as a voltage and power amplifier, since the input impedance is low while the output impedance is very high.

To obtain the best characteristics as an amplifier, it is essential to have proper coupling and source impedance. It is instructive to measure the gain as a function of frequency for a simplified transistor amplifier as shown in Fig. 18-55.

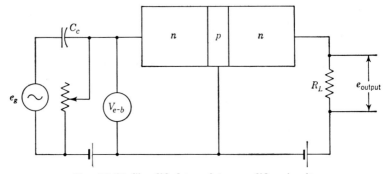

FIG. 18-55. Simplified transistor amplifier circuit.

18-51. Other Transistor Configurations. Transistors of the p-n-p type are also in common use. The basic operating principles can be discussed in the same manner as has been done for the n-p-n type. They differ in that the polarities of all batteries are reversed and, of course, the direction of current flow is likewise reversed.

18-52. Photoconductivity. In referring to the energy band system of semiconductors (Fig. 18-45), it is seen that the conductivity of these elements will depend upon the population of electrons in the conduction band. The promotion of electrons to this band by means of thermal agitation has already been discussed. The conductivity may, of course, be influenced by other methods which are capable of increasing the population in the conduction band. The most important of these additional methods is due to the photoeffect. If a light quantum has an energy $\geq h\nu$, the energy gap between the valence and conduction bands, it is capable of raising electrons into the conducting state and therefore increasing the conductivity of the sample. This can happen not only with pure semiconductors but also with doped samples and at diode or transistor junctions. This effect is quite widely used in the measurement of light intensities in various fields of spectroscopy.

REFERENCES

Albert, Arthur L.: "Electronics and Electron Devices," The Macmillan Company, New York, 1956.

Seely, Samuel: "Radio Electronics," McGraw-Hill Book Company, Inc., New York, 1956.

Seely, Samuel: "Electronic Engineering," McGraw-Hill Book Company, Inc., New York, 1956.

Terman, F. E.: "Electronic and Radio Engineering," McGraw-Hill Book Company, Inc., New York, 1955.

PROBLEMS

18-1. Compute the emission-current density for a pure tungsten filament and for an oxide-coated cathode at the following absolute temperatures: 800, 1000, 1600, and

2000°. Assume an empirical value of $A = 60$ amp/sq cm-°K² for tungsten and $A = 0.01$ amp/sq cm-°K² for the oxide-coated cathode.

18-2. Copper has a work function of 4.0 ev. Compute the maximum wavelength of radiation which is capable of freeing electrons from a copper surface.

18-3. A phototube, consisting of an evacuated glass envelope containing a photo-emitting surface of work function 1 ev and a collector, is bathed with a beam of intense light. The wavelength of the light is 6.6×10^{-5} cm. The photo emitter and the collector are connected externally by a 10-megohm resistance. Compute the current which flows.

18-4. A vacuum-tube diode is to be used as a half-wave rectifier with a 60-cycle voltage. The maximum of the a-c voltage is 200 volts. A load resistance of 1,000 ohms is placed in series with the tube. Plot the undistorted half of the sine wave and the actual voltage across the load resistance. Assume that for this tube structure the constant in Child's law is $\alpha = 2 \times 10^{-5}$.

18-5. Replace the vacuum diode of Prob. 18-4 by a thyratron or gas diode which has a firing voltage of 16 volts. Again plot the rectified wave form. (Assume that Child's law holds until the potential across the tube reaches the firing voltage; after the tube has fired, assume that its voltage drop remains constant.)

18-6. Consider ideal half-wave and full-wave rectifier circuits for which the peak of the unfiltered, rectified voltage is 100 volts in each case and the load is 500 ohms. The rectified voltage is to be smoothed by placing a condenser in parallel with the load. Estimate the energy which must be stored in such a condenser at the peak voltage so that the smoothed direct voltage will not drop below 90 per cent of the peak in each case. Do this for the following frequencies of the alternating voltage: 60, 200, 1,000, and 10,000 cycles per sec. What size capacitor should be used in each case?

18-7. A simple triode amplifier as shown in Fig. 18-19 is set up. The total capacitance from plate to ground (in parallel with R_L) is 10^{-11} farad. The amplification factor is 20, and r_p is 2,000 ohms. Compute the mid-frequency gain and the frequencies at which the gain has dropped to $1/\sqrt{2}$ times mid-frequency gain for the following values of the load resistor: 100, 1,000, 10^4, 5×10^4, and 10^6 ohms. Tabulate the product of the mid-frequency gain times the upper half-power frequency for each of the frequencies. (This is called the gain-bandwidth product.)

18-8. In the following circuit, plot the gain as a function of frequency, first with the inductance shorted out and second with the inductance in series with the load resistance. Pay particular attention to the region of the upper half-power point.

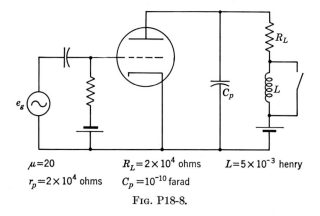

$\mu = 20$ $R_L = 2 \times 10^4$ ohms $L = 5 \times 10^{-3}$ henry

$r_p = 2 \times 10^4$ ohms $C_p = 10^{-10}$ farad

Fig. P18-8.

18-9. Calculate the voltage gain in the following circuit as a function of frequency.

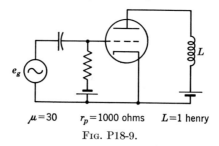

$$\mu=30 \qquad r_p=1000 \text{ ohms} \qquad L=1 \text{ henry}$$

FIG. P18-9.

18-10. Discuss the possible advantages and disadvantages of the circuit in Prob. 18-9 as compared with a similar amplifier with a resistive load.

18-11. When a condenser is charged from a constant voltage source V_0 through a resistance R, the potential across the condenser is

$$V = V_0[1 - \epsilon^{-(1/RC)t}]$$

Make a series expansion of the expression, and indicate the region in which it is valid to assume that the potential across the condenser rises linearly with time.

18-12. Compute the gain as a function of frequency for the circuit shown in Fig. 18-26 using the constant-current equivalent of a triode.

INDEX

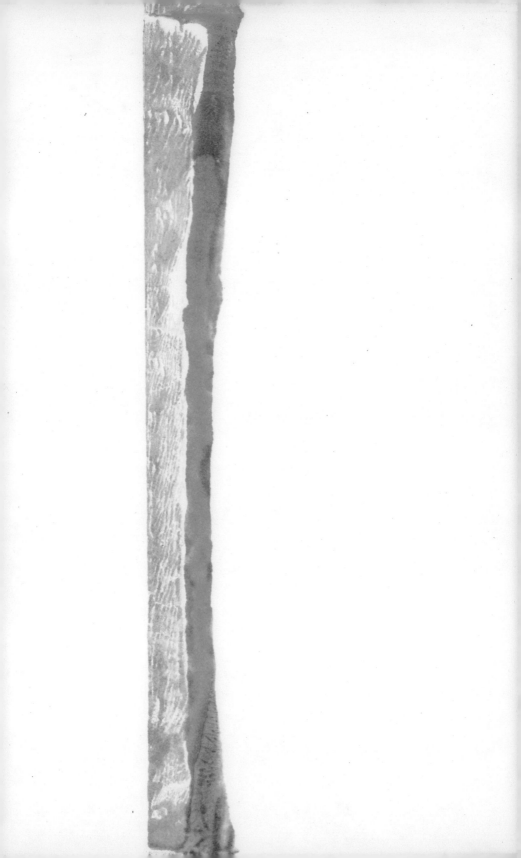